3 1299 00399 0968

P9-CFS-198

Also by Sharon Bertsch McGrayne

Nobel Prize Women in Science: Their Lives, Struggles, and Momentous Discoveries

365 Surprising Scientific Facts, Breakthroughs, and Discoveries

Sharon Bertsch McGrayne

John Wiley & Sons, Inc.

New York ► Chichester ► Brisbane ► Toronto ► Singapore

To George F., Ruth Ann, and Frederick M. Bertsch

This text is printed on acid-free paper.

Published by John Wiley & Sons, Inc.

Library of Congress Cataloging-in-Publication Data

McGrayne, Sharon Bertsch.

365 Surprising scientific facts, breakthroughs, and discoveries / Sharon Bertsch McGrayne.

p. cm.

Includes bibliographical references and index.

ISBN 0-471-57712-X (alk. paper)

1. Science—Miscellanea. 2. Engineering—Miscellanea. 3. Technology—Miscellanea. 4. Discoveries in science—Miscellanea. 5. Inventions—Miscellanea. I. Title: Three hundred sixty-five surprising scientific facts, breakthroughs, and discoveries.

Q173.M445 1994

500—dc20 93-38675

Printed in the United States of America.

10 9 8 7 6 5 4 3 2 1

► Illustration Credits

Page 3, Greeley and Hansen, Engineers; page 6, Mehmet Sarikaya and Ilhan A. Aksay, University of Washington; page 14, IBM Research Division; page 23, Fisons; page 31, Sierra Regional Eye and Tissue Bank, University of California Davis Medical Center; page 33, all, collection of James C. Whorton, M.D., University of Washington; page 48, Paul W. Sherman, Cornell University; page 66, John P. Dumbacher, University of Chicago; page 68, The Alex Foundation, University of Arizona; page 69, top, David B. Grobecker, University of Washington; page 69, bottom left, Theodore W. Pietsch, University of Washington; page 69, bottom right, Theodore W. Pietsch, University of Washington; page 71, Samuel D. Marshall, University of Tennessee; page 75, both, courtesy of Martin J. Muller; page 81, all, Babara A. Roy, University of California at Davis, reprinted with permission from Nature, vol. 362, cover and page 57; page 85, left, Fred E. Hillerman; page 85, right, Gene Kritsky, College of Mount St. Joseph and Cincinnati Museum of Natural History; page 98, Stanley M. Awramik, University of California at Santa Barbara; page 105, taken by Tony English, provided by William M. Denevan, University of Wisconsin at Madison; page 108, top, Bernard Hallet, University of Washington; page 108, bottom, Nancy C. Knight, National Center for Atmospheric Research/University Corporation for Atmospheric Research/National Science Foundation; page 117, Douglas C. Rees; page 118, left, Michael W. Day; page 118, right, Douglas C. Rees; page 125, left, Daniel Ugarte; page 125, right, Walter A. de Heer and Daniel Ugarte; page 133, both, The Nobel Foundation; page 134, Grabber International; page 140, Scott Sandford and NASA/AMES; page 155, Robert L. Ingalls, University of Washington; page 160, IBM Research Division; page 162, Sandra Johnson; page 173, IBM Research Division; page 181, Chaos Group, University of Maryland; page 184, both, Benno Artmann; page 188, Michigan Sea Grant, University of Michigan.

▶ Contents

▶ Preface

Dear Readers,

Originally, this book aimed to be simply a light and amusing collection of startling curiosities about science. The idea was to highlight odd and unexpected facts gleaned from hither and yon—the kind of juicy lore that someone might enjoy relaying to a friend over dinner.

It soon became obvious, however, that many of the most fabulous and unexpected tidbits were coming hot-off-the-griddle from scientific research. In trying to understand the natural and physical world, scientists were uncovering more surprises than anyone could ever have imagined.

So the book took on a new cast. It still aspires to be an entertaining book of scientific curiosities. But most of the facts are drawn from recent scientific studies. This is not to say that the research included is frivolous. Quite the contrary. It's the significant, "nontrivial" science that is turning up the surprises, challenging the imagination, and toppling the old myths.

The message—if there can be a message in something so trivial as a science trivia book—is that modern science is truly interdisciplinary. Engineers today study biological structures; chemists work in archaeological museums; physicists analyze AIDS epidemiology; and mathematicians use computers to run "experiments." Mathematicians, physicists, physicians, anthropologists, and chemists have flocked to genetics, for example, and made it one of biology's most rapidly developing areas.

Thoroughly plowed fields of research are harvesting surprises too. Chemists, physicists, and engineers who thought they knew everything about carbon are fascinated by the irresistible properties of an entirely new kind of carbon molecule. Thanks to new medical technology, botanists discover two plants. Exotic rarities? No, they are thought to be the two most common plants on our planet. Physiologists, ecologists, and anthropologists study how animals regulate the sex of their offspring and why some primates commit infanticide. Geneticists discover hereditary mechanisms that Mendel never dreamed of.

As scientific questions and evidence become more complex, researchers have tried to simplify their problems by studying nature at its most exotic and extreme: how life-forms cope with arctic cold; how matter behaves close to absolute zero; how marine animals survive enormous atmospheric pressures underwater; and how bacteria can thrive in boiling water.

But too much message can spoil the fun. And first and foremost, this book aims to entertain.

For readers who insist on learning more, suggestions for further reading appear at the back of the book. They are number-coded to correspond to each startling fact, discovery, or breakthrough and should be available in most city or college libraries.

Fortunately for science buffs, general-interest science magazines are published today for every level of sophistication. Among the least demanding are *Popular Science, Discover,* and *Harvard Health Newsletter.* The next tier up in difficulty includes *Science News,* which does a magnificent job of explaining science to nonspecialists in an interesting manner; *New Scientist,* a lively British magazine; *The New York Times* science pages, which appear on Tuesdays; and *The Sciences,* published by the New York Academy of Sciences. These publications cover all fields of science, mathematics, engineering, and technology. *Natural History* is at roughly the same level of difficulty but, obviously, covers only the natural sciences. *Tech Review,* a publication of the Massachusetts Institute of Technology, specializes in engineering and technology. C&EN, from the American Chemical Society, covers chemistry and chemical engineering.

Somewhat more technical are the articles in *Scientific American.* Sadly, the National Science Foundation ceased publication of its excellent magazine *Mosaic.* General commentaries, as well as technical journals, appear in *Nature, Science,* and *JAMA* (as the journal of the American Medical Association is now called).

While I hoped that this book would be entertaining fun, I also wanted it to be as accurate as possible. Thus, I'm especially grateful to the scientists who read portions of the manuscript. They include George, Fred, and Ruth Ann Bertsch, Diane duCros, Harold W. Kuhn, Arjen Lenstra, Timothy McPhillips, Tom Milac, Jeffrey S. Morgan, Scott A. Sandford, Margaret Thouless, James C. Whorton, and Alycia Weinberg. Reference librarians at the Seattle Public Library and the University of Washington libraries were especially helpful. And, as always, thanks to my agent, Julian Bach.

Cheers and have fun,

Sharon Bertsch McGrayne
February 1994

Engineering and Technology

▶1 **The fax took over the communications world in five quick years during the 1980s. Providing an easy, instantaneous way to transmit an image *and* a written record, it proved irresistible. In the United States, sales of facsimile machines doubled during the boom years of 1986, 1987, and 1988. When was the first fax patent issued?**

In 1843, 30 years *before* the telephone. British clockmaker Alexander Bain patented the basic concept of sending an image electrically. He used metal blocks to form the message; when a stylus moved up and down over the blocks, it broke and restored an electric current. In 1865, Abbé Caselli introduced the first commercial facsimile system between Paris and Lyons. Thomas Edison played with the concept, and newspapers began sending photographs in 1902. During World War II, both German and Allied forces faxed maps, photographs, and weather information. The Japanese developed modern fax machines because their written language—based on more than 2,000 characters—is ill-adapted to telegraph and telex systems.

▶2 **Three places in North America lie below sea level. Two are in the California desert. Where is the third?**

New Orleans. A bowl of water surrounded by water, the city sits between the Mississippi River and Lake Pontchartrain. As if this weren't water enough, more rain falls on New Orleans than on any other American city. A system of massive screw pumps built early in the twentieth century moves rainwater over levees to the Gulf of Mexico. Without them, the city would flood.

▶3 **Chicago-area brides and grooms beg to be married at this waterfall park. What's so surprising about that?**

It's a wastewater treatment plant. Despite improvements to the area's storm drains and sewage plants, the placid Calumet Waterway starved for oxygen each summer. So Chicago engineers mimic nature's mountain streams: They bubble the water over man-made waterfalls at five locations. At each one, canal water is diverted to a sidestream during summer months. There it is pumped to the top of a waterfall. The system adds some 25 tons of oxygen to the canal each day. And with landscaping, the Chicago area gained five popular parks and wedding sites.

▶4 **During the early 1980s, American athletes earned more Olympic medals in low-tech sports like running and swimming than in high-tech sports like bicycling and kayaking. The U.S. Olympic Committee formed a Sports and Equipment and Technology Committee to research the situation. As a result, new materials and designs are revolutionizing competitive sports. Match the technology to the sport in the following lists.**

Technology

1. **Spring aluminum alloys.**
2. **Carbon composites and foam rubber pits.**
3. **Kevlar and maraging steel (a jet fighter alloy).**
4. **Hollow aluminum .006 inch thick and syntactic foam (tiny glass beads in rigid foam).**

Sport

A. Cycling.
B. Diving.
C. Kayaking.
D. Pole vaulting.
E. Fencing.

5. Clockspring steel.
6. Spoon-shaped wings.
7. Recliners, disks, silicon-ribbed suits, and fairings.

F. Archery.
G. Weight lifting.

1-B. Modern diving boards have 15 percent more lift than those of the 1960s, and their lightweight tips minimize injuries, especially to the head.

2-D. When vaulters stopped using bambooo poles and sawdust pits, world records jumped from 15 to 20 feet.

3-E. Maraging steel is less brittle than conventional carbon steel so épée tips do not break off and injure fencers.

4-F. Syntactic foam forms the core of archery bows, while aluminum tubes make lightweight arrows.

5-G. Weight lifters use barbells on highly flexible steel bars. After raising the barbells to the chest, they wait momentarily until the whipping motion of steel helps them complete the lift.

6-C. Wing paddles give kayakers lift.

7-A. Racing bicycles are still Model Ts; Olympics committees have resisted improvements that would make cycling equipment too expensive for the public. Ninety percent of a cyclist's energy is spent fighting air resistance; thus ribbed clothing directs air over the cyclist's body, and fairings (smooth wraparound structures) reduce air turbulence behind the rider.

▶5 **A famous, high-tech maker of scientific instruments commissioned a study of the future in 1967. The report predicted that by the year 2067 Americans would live in domed cities and watch three-dimensional TV. The report *failed* to predict that within five years the company's main product would be totally obsolete. What did the company manufacture?**

Slide rules. Keuffel & Esser was the company. Texas Instruments began making pocket calculators in 1967. By 1976, Keuffel & Esser was selling more Texas Instrument calculators than slide rules, and slide rules accounted for only 5 percent of its sales. The company mothballed its manufacturing equipment and sent it to the Smithsonian Institution as a piece of Americana.

▶6

Serendipity is the art of making an accidental discovery while looking for something else. Match the discovery with the subject the scientist was studying at the time.

The Discovery

1. **Liquid crystals.**
2. **Teflon.**
3. **The fact that auto exhaust pollutes the air.**
4. **Cortisone, the rheumatoid arthritis drug.**
5. **Cisplatin, the first heavy metal anticancer drug.**

The Subject under Study

A. **The effect of electric fields on cell division.**
B. **Refrigerants.**
C. **Botany.**
D. **Volatile oils in plants.**
E. **Adrenal gland.**

1-C. 2-B. 3-D. 4-E. 5-A.

Basic research is one of the raw materials required for serendipitous discoveries. A study of 76 major corporations in seven industries found that 10 percent of their new products and processes were based on fundamental research conducted within the previous 15 years. Fundamental research tries to establish new scientific facts, whereas applied research uses previously known facts to solve practical problems. Without fundamental research, the companies would have lost 5 percent of their total sales, or $41 billion a year.

▶7

Why is slug mucus like a liquid crystal?

Slug mucus may be sticky, slimy, and socially unacceptable, but sometimes it's a high-tech, liquid crystalline fluid. With its rod-shaped molecules aligned in one direction, slug mucus is midway between a crystalline solid and a conventional liquid. The degree of molecular alignment may affect how viscous the mucus is.

The slug changes the viscosity of its mucus according to the ground it is crossing. Crawling over rough soil, it protects its body by producing heavy, goopy material. Edging across the smoothly painted wall of a house, it quickly switches to a thinner mucus. Like smart material, it responds to varying conditions in its environment.

Inside the slug, the molecules of mucus are stored in tightly packed, accordianlike bunches. Once the molecules are secreted, they are aligned parallel-fashion like liquid crystals. They soak up water and explode like a jack-in-the-box, expanding their volume up to 50 times in a fraction of a second.

Slugs may hold the key to a number of medical disorders involving mucus. In cystic fibrosis, heavy mucus clogs the respiratory system. Ulcers may occur when the mucus fails to act as a barrier protecting the gastrointestinal tract. Infertility can be caused by sperm's inability to swim through cervical mucus fast enough to reach a viable egg in time.

▶8 You can jump up and down on an abalone shell, but it doesn't break. Yet abalone is made of calcium carbonate, which is basically a brittle chalk. So why doesn't abalone break?

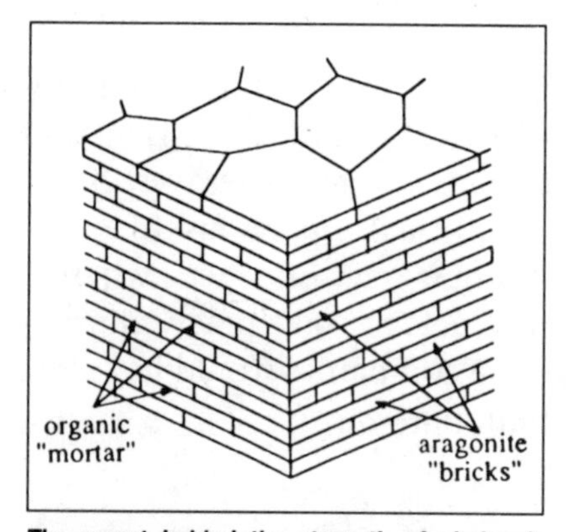

The secret behind the strength of abalone's mother-of-pearl, as revealed by a transmission electron microscope.

Because of its brick-and-mortar construction. The abalone, like the pearl oyster and the nautilus, builds a natural ceramic that is stronger and twice as tough as any high-tech synthetic ceramic made today. Six-sided calcium carbonate bricks are reinforced with a thin layer of fibrous mortar composed of proteins and carbohydratelike molecules. This mortar makes abalone shell extremely tough. Cracks in the shell do not grow as easily as those in most synthetic materials.

Ceramic polymer composites are being developed to mimic this natural composite material.

▶9 What was the most important source of power during the Industrial Revolution? *Hint:* The steam engine, invented by James Watt in the eighteenth century, ushered in the Industrial Revolution.

Water mills. Simple paddle wheels immersed in streams were used in China and the Middle East and introduced in Rome about 70 B.C. Late in the Empire, the Romans invented the overshot wheel where water enters the wheel at the top and pulls the wheel around. The Romans preferred to use slave labor, however. During the Middle Ages, water mills became an important source of power. They ground wheat, fulled woolen cloth, sawed timber, crushed ore, hammered metal, blew furnace bellows, tanned leather, made beer and paper, and pressed oil from seeds. The English Domesday Book of 1086 listed 6,000 water mills just for grinding grain. By the thirteenth century, France was dotted with thousands of water mills. During the Industrial Revolution, they were greatly improved, although they still averaged less than ten horsepower each.

▶10 What marvel of chemical engineering makes *all* the following products?

A. A five millisecond epoxy glue.
B. A water-soluble material produced in *water*.
C. A flexible, but inelastic fiber.
D. Another fiber stronger than nylon fishing line and ten times tougher than Kevlar. Only one-thousandth the thickness of a nickel, it can support an apple.

***Hint:* The manufacturing process requires little new material and takes place at normal, ambient temperatures. No harsh chemicals are required.**

Spiders. One spider may produce seven different kinds of silk with varying combinations of amino acids.

A. Spiders use fiber-reinforced epoxy to glue their radial web and drag lines to a support.
B. Although the silk is dissolved in water while being synthesized inside the spider, it is insoluble in water outside the spider. This feat has not been duplicated industrially.

   It is shear forces, not high temperatures, that make the silk water-insoluble. Inside the spider, the silk consists of protein molecules in a liquid crystalline state. Thanks to the common alignment of its rod-shaped molecules, the glop flows easily as the spider extrudes it through a tiny hole. Shearing forces stretch the molecules, changing both their shape and their bonds so that they bond together more easily than to water. Industry may try to adapt this technique to make synthetic fibers.
C. The spider uses flexible silk to build its radial web lines and to wrap its prey.
D. For a combination of strength, stiffness, and toughness, the spider's drag line outperforms any synthetic polymer.

Oddly enough, what comes out of the spider does not depend on what goes into it. A spider needs only a little additional protein to make its web. It simply eats its old web and, 30 minutes later, spins a new one. A spider is the epitome of biodegradable recycling.

▸11 Match the following famous flops with the reasons for their failure.

The Flop

1. Kresge Auditorium, Massachusetts Institute of Technology.
2. John Hancock Tower, Boston.
3. The Tacoma Narrows Bridge.

The Reason

A. Several thousand window panes popped out.
B. Only four months old, it rippled and writhed in a moderate wind.
C. Its elegant lead and concrete roof cracked and flooded.

1-C. The roof of the Kresge Auditorium, designed by famous Finnish architect Eero Saarinen in 1955, was a goof. Its triangular, concrete curves were designed for looks, not leaks. During every heavy rainstorm, cracks in the concrete flooded the auditorium. Twenty-four years later, the building had to be reconstructed.

2-A. One quarter of the 10,300 glass panels in I. M. Pei and Partner's 60-story Hancock Tower fell to the ground in downtown Boston between 1971 and 1973.

3-C. The Tacoma Narrows Bridge in Tacoma, Washington, collapsed into the water in a mild windstorm on November 7, 1942.

▸12 Skintight suits, teardrop helmets, shaved legs, and tights—how much advantage do they give a bicycle racer?

Six to 10 percent. At speeds above 20 miles per hour, 90 percent of a biker's energy is devoted to overcoming air resistance.

▶13 Auto racing on public roads was illegal in the United States during the 1920s, yet during those years American cars established speed records that stood for decades. How?

To circumvent the law, promoters built wooden racetracks as steep as salad bowls 2 miles around. Millions of board feet and literally tons of nails were used to build each track. As smooth as dance floors, they were far superior to any road at the time. The steepest of America's 24 wooden speedways was banked at 50 degrees—so steep that a car moving at less than 110 miles per hour slid down. Most of the racetracks lasted for only two or three years, and none survived the Depression.

As a result of these wooden tracks, American racing cars were designed differently from everyday, road vehicles. Each racer tended to specialize in one particular function, like straight-line speed, left-hand turns, or speeds above 200 miles per hour. In Europe, general-purpose vehicles raced on public roads, and improvements to racing cars benefited consumers' vehicles too.

▶14 One day the spade of an English peat cutter struck pay dirt—a plank of hard wood. Digging farther afield, archaeologists discovered a mile-long ditch filled with wooden planks. What had the peat cutter unearthed?

The world's oldest road. Neolithic settlers about 6,000 years ago built what is now called The Sweet Track, after peat cutter Raymond Sweet. Using sharpened stone tools, skillful roadbuilders hewed wooden planks from 400- and 120-year-old oaks that had grown in carefully managed forests. The footpath, which was only 12 inches wide, was used for about a decade. Then it flooded and was covered over by 30 feet of waterlogged peat. The acidity of the water protected the wood from bacteria, fungi, and desiccation.

▶15 What is the world's smallest battery?

A device 100 nanometers square, each nanometer being one-*billionth* of a meter long. The battery was made at the Massachusetts Institute of Technology. Ultrasmall fuel cells will produce a few milliwatts of power per square centimeter. Tiny vacuum tubes will be one-hundredth the thickness of a human hair. Nanotechnology uses electron beams, X rays, deep ultraviolet light, or ions to chisel patterns into a resist, which is transferred to a material like silicon or gallium arsenide.

▶16 **Early in the twentieth century, two neighboring farms were cleared in the state of Washington. Although each started out with the same soil, one neighbor ended up dirt-poor and the other ended up dirt-rich. The soil of the dirt-rich farmer is now filled with microorganisms and enzymes; it contains 60 percent more decayed organic matter, more water, and more nutrients. It is softer, so seedlings emerge more easily. And the layer of his fertile topsoil is 16 centimeters thicker than his neighbor's, which is eroding rapidly. Why is one farm so much healthier than its neighbor?**

The neighbors used different farming techniques. They operated identically until after World War II, when cheap fertilizers and pesticides became available. Then the neighbors had to choose. One learned new, high-tech methods dependent on agrochemicals. The other continued with traditional, organic methods. He grew green manure crops, plowing them over one year out of every three or four years.

Who was right? Judging by the quality and quantity of their soil today, the stick-in-the-mud farmer was smart. He's dirt-rich and will be for years to come. It's his modern neighbor who is dirt-poor.

Their farms lie in one of the world's richest wheat and pea growing areas. Its hilly terrain and its reliance on high-tech farming also make it one of the most rapidly eroding regions in the United States.

▶17 **The Channel Tunnel between England and France is actually two 50-kilometer (30-mile) long tunnels, connected by cross passages. Why are the passages between the two tunnels necessary?**

To reduce aerodynamic drag. Pushing a column of air the entire length of the tunnel would slow down any train.

▶18 **Which came first: horseback riding or the wheel?**

Horseback riding may beat the wheel by 500 years. If so, riding horses was the world's first significant innovation in land transport. Riding began in the Ukraine about 6,000 years ago, if dental wear caused by a bit in the mouth of a prehistoric horse can be trusted. If so, the skeleton of a seven- or eight- year-old stallion found beside two dogs in the Ukraine is the oldest known mount in the world. Their burial site dates from the Ukrainian Copper Age around 4000 B.C. Analysis by a scanning electron microscope revealed that the horse's teeth had been worn down by a bit, indicating that a rider had controlled the horse's movements from behind. Until then, horseback riding was thought to have begun in central Asia about 1500 B.C. The earliest evidence for wheels dates from about 3300 to 3100 B.C. in southern Poland, Hungary, the Ukraine, the Russian steppes near the Black Sea, and Mesopotamia.

The new evidence supports an old theory: that horseback riders from the Eurasian steppes helped spread Indo-European languages, the most widely dispersed language family in the world.

▶19 Buffalo were the predominant large grazing animals in North America. What creature filled the same niche in the Eurasian steppes as it roamed about in herds?

Wild horses. By 1800 B.C., domesticated horses were pulling war chariots in the Middle East.

▶20 The Great Chicago Fire of 1871 was succeeded by the Great Chicago ________ of 1992. (Fill in the blank.)

Flood. An abandoned, 80-year-old system of freight tunnels under downtown Chicago sprang a leak on April 13, 1992. River water flooded the 50-mile system and the basements of more than 100 buildings. Floodwaters closed adjacent subway tunnels and freeway lanes, shut off public utilities in Chicago's business district, and threatened the foundations of landmarks like the Chicago Art Institute and Marshall Field's department store. To plug the leak and to drain downtown, engineers drilled six access shafts into the tunnel; filled them with sand, rock, and cement; and pumped out 130 million gallons of water—all within the next 74 days.

▶21 **Even if you love flying, you should get a sinking feeling at this airport. Its control tower and eight-story administration building are settling 11 to 13 meters into a bay. They were built above a floating foundation on a human-made, sinking island. If the buildings do not settle simultaneously, they and the pedestrian bridges connecting them will break. An elaborate system of hydraulic jacks and computers monitors and corrects for differential settling. Where is this floating/sinking/wobbling system located?**

Five kilometers (3 miles) off Osaka, Japan, at the New Kansai International Airport. It is Japan's first 24-hour-a-day airport. The island it sits on was constructed with fill 33 meters deep. But the weight of the fill is expected to compress the clays underneath by 11 to 13 meters over the next 30 years.

▶22 **When does a trash heap become a power station?**

When its methane gas is tapped. Approximately 100 landfills in the United States collect the methane gas they produce. Then they burn the methane, which is the primary constituent of natural gas, to turn turbines and produce electricity and/or steam. America's largest landfill power station produces enough electricity for 100,000 homes. Located in Whittier, California, it has been in operation since 1986.

Four hundred vertical wells, trenches around the perimeter of the landfill, and 40 miles of pipe collect the gas. Methane does not burn in open air, but it can if trapped in an enclosed space, like a basement. Landfill gas is cheaper than natural gas. Collecting and burning it also prevents its escape into the atmosphere, where it is a major contributor to global warming.

Landfill power stations in the United States consume approximately 1,000 metric tons of methane annually and produce about 787 megawatts of electricity.

▶23 **Believe it or not, water has become a high-tech, high-precision tool for cutting through metal, rocks, ceramics, plastics, rubber, food, leather, and paper. As a machine tool, water can turn out complicated patterns with cuts only 0.1 to 0.8 millimeters wide. What makes cheap, clean, safe, and accessible water so high-tech?**

Supersonic speed. The water jets out at 540 to 1,400 meters a second—significantly faster than the speed of sound through air. Speeding through a fine nozzle only 0.05 to 0.5 millimeter in diameter, water breaks microscopic particles off materials. Plain water cuts food, leather, and paper products. With abrasives and other additives, it slices metal and other hard materials. Use of supersonic water jets eliminates mechanical damage to the surfaces being treated.

▶24 **What engineer:**

- **Built the highest arched bridge and some of the longest span bridges in the world?**
- **Prefabricated churches for third world countries and easy-to-assemble bridges for the military?**
- **Engineered what remained the world's tallest structure for more than 40 years? When built, it was double the height of the next-tallest building.**

Alexandre Gustave Eiffel, designer of the 1,000-foot-tall Eiffel Tower in Paris. A leading nineteenth-century engineer, Eiffel specialized in wrought iron, innovative construction methods, and careful wind bracing. Parts for the Eiffel Tower were engineered to be accurate to within one-tenth of a millimeter. Until then, the tallest building had been the Washington Monument in Washington, D.C.; it was only 555 feet tall. The Eiffel Tower remained the world's tallest structure until the Chrysler Building opened in 1930 in New York City. Most of Eiffel's bridges and other structures are still in use.

▶25 **A nanometer—a billionth of a meter—is to a pebble as a pebble is to:**

A. The height of the Sears Tower in Chicago.
B. The distance from New York City to San Francisco.
C. Earth's diameter.

C, Earth's diameter. Even 100 nanometers is incomprehensibly small: the distance that a human hair grows in ten seconds. In 1960, the transistors in early integrated circuits had features approximately one-third the diameter of a human hair. Today, advanced commercial transistors are 100 times smaller than a hair. Nanotechnology involves features 3,000 times smaller than a hair—a little more than a hundred atoms across.

In conventional transistors, electrons behave like particles. But in nanoengineering, dimensions are comparable to electron wavelengths. And when electrons are confined to regions the size of their own wavelengths, they behave like waves.

▶26 **Big corporations normally want big logos. They fly them from blimps, put them in lights over Times Square, and paste them on billboards over freeways. What company wanted the world's *smallest* logo?**

IBM. In 1990, Big Blue built its logo 13 atoms high—a mere five-billionths of a meter from top to bottom. Its researchers used the tip of a scanning tunneling microscope to pick up one atom at a time and place it in position on a crystal of nickel. As an exercise in nanotechnology, the "I" was composed of nine atoms of xenon. The "B" and "M" were bigger: Each used 13 atoms. But the letters were still only five nanometers or five-billionths of a meter tall.

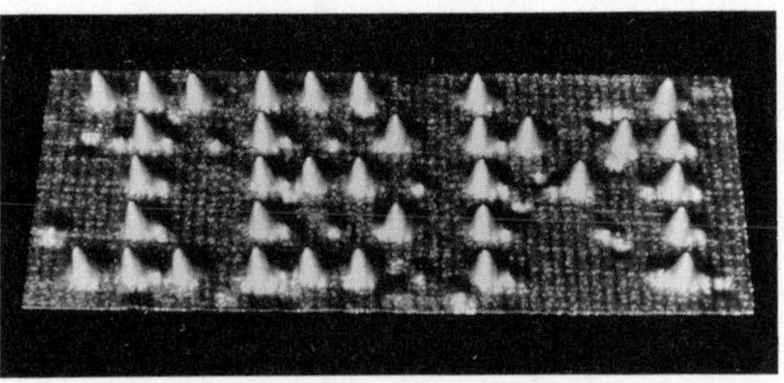

A corporate logo 13 xenon atoms high and 660 billionths of a meter wide. The atoms are about 50 billionths of a meter apart. They were positioned and imaged by a scanning tunneling microscope.

By the year 2000, engineers hope to be mass-producing transistors 25 nanometers small. At the scale of 0.1 nanometer, the individual atoms that form chemical bonds become identifiable.

Don't hold your breath waiting for nanometer-sized advertising, though. When the temperature rose above −243 °C (−380° F), "IBM" fell apart.

▶27 **How far does light travel in one nanosecond?**

A. 1 foot.
B. 1 meter.
C. Several kilometers.

A, 1 foot. A foot is a rather short distance, but then again, a nanosecond is an infinitesimal amount of time: 10^{-9} second or 0.000000001 second. As a matter of fact, a nanosecond is to a second as a second is to 30 years. Althogh a nanosecond may seem short, many scientists today study phenomena that are a thousand times shorter.

▶28 **Gene pharming for fun and pharmaceuticals may be the wave of the future. What gets pharmed?**

A. Cows.
B. Sheep.
C. Goats.
D. Rabbits.
E. Pigs.

All are prime candidates for pharming. In gene pharming, farm animals are used to produce medicines—therapeutic proteins—in their milk or blood. Domesticated animals like cows, sheep, and goats have been milked since neolithic times. And rabbits and pigs produce large quantities of milk relative to their size for their big litters. Three Scottish sheep have already produced a protein, the lack of which kills people with inherited emphysema. Human plasma, the current source, cannot meet world demand for the protein.

A human gene that makes a desirable protein is fused to a section of animal DNA that controls protein production in the mammary gland. Then the hybrid gene is injected into early animal embryos and transplanted into a foster mother animal. So far, though, too few of these transgenic embryos make it to babyhood, much less motherhood.

▶29 **Like French fries and salad dressings, hydraulics systems are switching to _______. (Fill in the blank.)**

Vegetable oils. In the case of hydraulic fluids, rapeseed oil is the favorite. Environmentally acceptable hydraulic fluids are being used in agriculture, forestry, and other sensitive applications so that spills and leaks don't harm groundwater and animal life. Readily biodegradable and nontoxic, vegetable oils lubricate hydraulic pumps and valves.

▶30 **One hot summer day in Texas, a construction crew was laying a pavement of exposed aggregate. When they finished, the foreman exclaimed, "Hey, it's time for lunch!" Popping open a can of Coca-Cola, he sprayed the concrete surface with the liquid. Then he took off for lunch. What had the foreman done wrong?**

He'd used Diet Coke. Sugar retards the hardening of concrete, but aspartame is not a sugar and will not do the job.

When a ready-mix concrete truck breaks down and its load hardens, its driver is often held responsible for jackhammering the concrete out of the truck. So, many drivers haul around a few bags of sugar in addition to their regular load. Sweet solution, eh?

▶31 **Who built the longest interstate highway system—the Roman Empire or the United States?**

The Romans. They built almost 53,000 miles of roads, compared to the U.S. interstate highway system, which in 1992 extended only 42,795 miles.

▶32 **Roman roads set construction standards for 2,000 years. Their pavement was between 3 and 5 feet thick. Their surfaces were sloped for drainage and incorporated ditches and/or underground drains. They are so tough that Israeli tanks used one to cross the Negev and chase the Egyptian army into the Sinai during the 1948 War for Independence. Why can't modern engineers build roads that last that long?**

Cost. Roman roads were extremely expensive. Even using slave and conscript labor, the Roman military spent about $3 million (in 1993 U.S. dollars) per mile to build the Appian Way in Italy. And by today's standards, the road is too noisy, rough, and narrow. Intended for use by foot soldiers, a Roman road could never accommodate today's traffic volume.

▶33 **Road wear-and-tear is heavier today than 200 years ago because of:**

A. Greater vehicle weight per wheel.
B. More vehicles.
C. Poorer construction.

B, more vehicles. Tire loads today are roughly the same or lighter than they were in 1809. When fully loaded, a lightweight wagon pulled by four horses weighed 3.5 tons. With 3-inch wide wheels, the wagon weighed 1,960 pounds per wheel or 653 pounds for each inch width of tire. Today, permissible tire loads vary by state from 500 to 800 pounds per inch width. The difference in road usage is thus not weight per wheel inch but the enormous numbers of vehicles on the road. Today, a quarter of a million vehicles may speed through a busy federal highway interchange each day.

▶34 **How did archaeologists find:**

- **A farmer's field mulched 550 years ago with pebbles to conserve water?**
- **Thousand-year-old footpaths around a Costa Rican volcano?**
- **A network of trade routes linking Chaco Canyon in New Mexico to surrounding Indian villages between A.D. 900 and 1100?**

By using noninvasive, remote sensing technology instead of pick-and-shovel digging. Aerial photography using near-infrared wavelengths reveals stresses to plant life that the eye cannot see. Thanks to their densely compressed soil, thousand-year-old footpaths near the Arenal volcano become visible where they bend around a supply of funerary rocks to head down to a stream. Heat-detecting infrared scanners and ground-penetrating radar also help archaeologists avoid digs, which can destroy the very remains they want to study.

▶35 **What is the region of the United States with more Federal Superfund sites than any other area its size?**

California's Silicon Valley. About 150 toxic sites were monitored in 1993. One in five was considered so serious that it was put in the Federal Superfund for cleanup. The computer chip industry had prided itself on being squeaky-clean because it did not cause air pollution. But codes regulating underground building tanks for solvents were not strict enough.

▶36 **Ray Turner, a production engineer in a circuit-board factory, was chewed out for leaving a tank of ozone-depleting chlorofluorocar-**

bons (CFCs) wide open. Computer chip manufacturers used CFCs to remove pine resins from printed circuit boards after soldering. Turner decided to eliminate his problem by eliminating CFCs.

Tinkering at his kitchen sink, Turner produced Turner's Crazy Flux. Turner's Crazy Flux is cheaper than CFCs, more effective, and environmentally safe. It can be cleaned off with water. And the only retooling required for the production line was a dishwasher. What was Turner's Crazy Flux?

Lemon juice. Turner turned to lemon juice after trying vinegar and lemon rind. To help eliminate ozone depleters, other companies opt for extract of orange rind, plain soap and water, and a chemical that occurs naturally in peaches, plums, cantaloupes, and other melons.

▶37 **What do the Negev of Israel, the wadis of Libya, the Thar Desert of western India, the Papago and Anasazi Indians of southwestern United States, and the *qanats* of Iran have in common?**

Ancient water-gathering systems that made farming flourish in the desert. During storms, scarce rainfall was collected from the hillsides and concentrated by stone walls, dikes, or tunnels into cisterns or fields below. Water "harvesting" made Libya the granary of the Roman Empire 2,000 years ago. Nabataean waterworks, reconstructed after 1,500 years, enable Israeli farmers to grow crops where rainfall is only 4 inches a year. In Southwest Asia, miles of tunnels were built 2,000 years ago in hillsides to collect water from its source. Qanats, maintained to modern times by serfs and children, are still the traditional source of water through hilly southwestern Asia. Iran has an estimated 40,000 tunnels extending 160,000 miles.

▶38 **What is the most densely populated desert in the world?**

The Thar Desert of western India, where ancient water-collecting systems capture enough water to permit 60 inhabitants per square kilometer. The system traps seasonal rains in tanks, cisterns, walls, dams, and artificial water holes.

▶39 **A film showing the collapse of the Tacoma Narrows Bridge in 1942 is a favorite among physicists. It shows what can happen when a sys-**

tem is disturbed by external forces at the same frequency as its own natural oscillations. Forty-seven years later a similar accident occurred during an earthquake. Where did that happen?

On the Nimitz Freeway in Oakland, California, on October 19, 1989. During the earthquake, the elevated highway warped in waves and collapsed, crushing at least 38 motorists. Scientists found evidence that the ground had been shaking during the quake at the same frequency at which the highway naturally vibrated. Weak brick and masonry buildings next to the highway were unharmed.

Medicine, Health, and Nutrition

▸40 **When did AIDS develop?**

A. 1981.
B. 1959.
C. 100 to 200 years ago.

A, B, C. The first indication that AIDS would become an epidemic occurred about 1981. But a Manchester sailor died of AIDS in 1959. And mathematical models indicate that the virus could have slowly spread unrecognized in parts of Africa for 100 to 200 years and perhaps even longer.

▸41 **For 50 years, patients have been told that stress, diet, smoking, and drinking cause ulcers. In fact, ulcers have come to symbolize the harried executive, and the expression "You give me an ulcer" means that you are making me nervous. But now the phrase "You give me an ulcer" means something else. What?**

That most ulcers are infectious and caused by infectious bacteria, not stress. Antibiotics kill the bacteria and provide a permanent cure in about 80 percent of patients. In contrast, the conventional treatment—antacids or acid-lowering drugs—has a recurrence rate of about

95 percent within two years. The bacteria *Helicobacter pylori* survive the stomach's acidity thanks to an unusual enzyme. The enzyme actually decreases acidity in the stomach by breaking down urea into carbon dioxide and ammonia. A childhood vaccine against the bacteria might eliminate many ulcers. *Helicobacter pylori* may also be responsible for chronic atrophic gastritis, a precursor of gastric cancer.

▶42 Which public health problem kills more Americans than any other?

A. AIDS.
B. Alcohol.
C. Car accidents.
D. Crack cocaine.
E. Fire.
F. Murder.
G. Smoking.

G, smoking. It kills more Americans than *all* other causes of death combined. One thousand Americans die daily from the effects of smoking.

▶43 What disease has killed more Americans and Europeans than any other in recorded history?

Tuberculosis, still the leading cause of worldwide deaths from infectious diseases. In addition, the emergence of drug-resistant strains threatens the ability of modern medicine to control the diease.

▶44 Why sing of mites, mattresses, and making love?

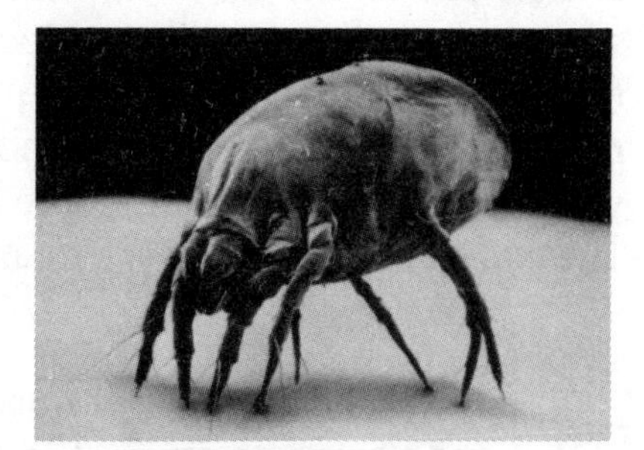

A dust mite, magnified 300 times in this scanning electron microscope image.

Because semen, it seems, provides a high-protein diet for the dust mites that inhabit bedding and make allergic lovers sneeze and wheeze.

Allergies to dust are actually allergies to the virtually invisible mites that feed on human skin scales. Most asthmatics, for example, are allergic to dust mites. The insects love to nest in rugs, upholstery, bedding, and mattresses. In particular, mites thrive in lovers' beds. Mites in semen-spotted bedding produce more eggs than mites that eat only dead skin scales.

Condoms, anyone?

▶45 In a remote region of Papua New Guinea, the incidence of asthma soared almost 40-fold in ten years. What was the culprit?

Cotton blankets, which had just become popular for use at night. Bedding is a prime breeding ground for house dust mites, microscopic insects that eat dust specks and dead skin cells. Dust allergies are actually allergies to these mites. They nest in rugs, upholstery, and bedding. Two million dust mites can call one double bed their home.

Modern air-tight home construction protects house mites, which is one reason why both the incidence of childhood asthma and death rates from asthma have increased rapidly. During the 1980s, children's death rates from asthma nearly doubled, although the death rate from virtually every other childhood disease except AIDS dropped.

▶46 NO news is good news—for whom?

Impotent men, most of whom cannot have sexual intercourse because of problems with blood flow in the penis.

NO—nitric oxide—is the chemical signal that brings about an erection. Nitric oxide, *not* the dental anesthetic nitrous oxide, relaxes the smooth muscle in two long, expandable sacs of the penis. As a result, arteries can fill the sacs with blood. As the inner spongy walls of the sacs expand, they prevent the blood from draining out. Trapped inside the penis, the blood makes the penis erect. An estimated 10 million men in the United States are impotent. Whereas only 2 percent of 40-year-old men are impotent, more than 25 percent of 65-year-olds are. The condition is most prevalent among diabetics.

▶47 The care and feeding of a beer belly requires more than just alcohol and a couch-potato lifestyle. What is the third ingredient?

Eating fat. The body hoards fat, especially in the presence of alcohol. A person who consumes alcohol *and* fat burns the fat extra slowly.

And the excess, unburned fat gets stored, especially on the thighs and belly. In an experiment, men who consumed three ounces of pure alcohol a day—about six shots of whiskey or six beers—burned one-third fewer fat calories. People who eat carbohydrates—sugar or starch—instead of fat tend to burn most of their calories.

►48 **Reading the classics is sometimes complicated by their use of outdated names for diseases. When the heroine has dropsy, for example, does she have an itch or a fatal illness? When the hero has the lues, is he a roué or an innocent? Match these old and new names of diseases:**

Old	*New*
1. Ague.	**A. Stroke.**
2. Lues.	**B. Undulant fever.**
3. Apoplexy.	**C. Any wasting disease; TB.**
4. Dropsy.	**D. Malaria.**
5. Intermittent fever.	**E. Syphilis.**
6. Tabes.	**F. Edema (swelling caused by too much fluid in the tissues).**
7. Phthisis.	**G. Tuberculosis.**

1–D. 2–E. 3–A. 4–F. 5–B or D. 6–C. 7–G.

►49 **Feeling tense and cranky, tired and blue? Craving chocolate, sweets, and goo? What should a woman do when she feels depressed from premenstrual syndrome?**

Cram carbohydrates to increase the amount of serotonin in the brain. The chemical induces sleep and alleviates depression. Theoretically, according to experts, the carbohydrates can be eaten plain and unadulterated in potatoes, bread, rice, and pasta. Unfortunately, the effect lasts only a few hours and may have to be repeated fairly often. And even worse, considerable practical and anecdotal evidence suggests that carbohydrates may work best in cakes, cookies, chocolates, and chips. Theory may not always be an accurate reflection of reality. The ultimate answer for severe cases of premenstrual syndrome, experts retort, may be foods engineered to induce more serotonin with fewer calories.

▶50 How much of a woman's body is fat?

Between 26 and 28 percent (for the body of the average 18-year-old American woman—5 foot 5 inches tall and 126 pounds). In contrast, the body of a boy at the same height and weight is 15 years old and only about 12 percent fat. When men have finished growing, their bodies are still only about 15 percent fat.

The average woman stops menstruating when she diets or exercises so strenuously that she loses 10 to 15 percent of the normal weight for her height. At that point, she has lost about one-third of her body fat, and her hypothalamus, which puts out the signal that controls ovulation, becomes dysfunctional.

Women may need a store of easily mobilized energy to ovulate and menstruate. Pregnancy alone requires about 50,000 calories over and above normal metabolic requirements, and nursing consumes 500 to 1,000 calories daily. In prehistoric times, women who could stop menstruating when food was scarce and begin again when food returned reproduced successfully. In contrast, women who continued to ovulate when food was scarce did not reproduce successfully and left no descendants.

▶51 The Spanish flu of 1918, the Asian flu of 1957, the Hong Kong flu of 1968, and the Russian flu of 1977—all were misnomers. Choose a more accurate name for these worldwide epidemic killers.

A. Peking duck flu.
B. Antigenic shift virus.
C. Integrated pig-duck farming flu.

Any of the above. Chinese ducks are the world's primary reservoir for influenza viruses. China has more ducks than people, a very large number indeed. And ducks, which live in close proximity with pigs and farmers there, are as likely to be infected with human or mammalian viruses as with the avian variety. Their intestines, and those of the pigs, are veritable mixing bowls of human and avian viruses. Because the genetic material of influenza viruses is loosely strung together, it gets dislodged easily and reattached to new strains inside the animals. Wild ducks, flying from place to place and defecating and drinking in farm ponds, spread the new forms.

▶52 **On the back of what animal did the Black Death start?**

The marmot. Trappers who collected the pelts of dying and dead marmots in the Central Asian steppes were among the first reported victims of the plague. The pelts, carried west to European markets along the trans-Asian Silk Road, spread the fleas that spread the disease from rodent to rodent and from rodent to people. From Black Sea ports, shipboard rats spread their plague-infested fleas rapidly through the Mediterranean and Europe. Bubonic plague killed 20 million Europeans within four years in the fourteenth century and returned periodically for the next 400 years. With every outbreak, the plague became less virulent. A new species of plague bacillus may have evolved gradually, vaccinating animals and people alike to the worst varieties of the disease.

▶53 **Why do some infants sometimes fuss and refuse to nurse after their mothers have gone jogging? Do the babies expect their mothers to stick around all day?**

Not necessarily. Mama may have left a sour taste. Lactic acid, a byproduct of heavy exercise, can sour a mother's milk for as long as 90 minutes after exercise.

▶54 **Imagine an illness that strikes 1 percent of the population. A test is discovered but it is only 95 percent reliable; it gives false positives to five out of every hundred people tested. If you were to take the test and the results came back positive, what's the chance that you really have the disease?**

One in six. If 100 people take the test, five of them will be correctly identified as suffering from the disease while five healthy people will get false alarms. Thus, screening for a disease that affects only a small fraction of the population is quite unreliable. With a disease affecting 50 out of 100 people, the five false alarms would seem less significant.

▶55 **More years could be added to the world's life expectancy by eliminating accidents than by eliminating all deaths from cancer, chronic lung diseases, pneumonia, and influenza. True or false?**

True. In 1989, accidents cost every 100,000 Americans more than a thousand years of potential life before the age of 65. In comparison, all cancers cost the same group only 843.3 years; pneumonia and influenza 84.9 years; and chronic obstructive pulmonary diseases such as bronchitis, emphysema, asthma, and tuberculosis 62.4 years. Accidents cause such a high number of years lost because more young people have accidents while more old people get cancer.

▶56 **What hobbies cause the most hearing damage?**

A. Loud rock music and headphones.
B. Gun shooting.
C. Power boats and snowmobiles.

In order of damage, B–gun shooting, C–power boats and snowmobiles, and A–rock music. The most common recreational cause of serious hearing loss in the United States is gunshot noise, such as rifle fire. Oddly, the more damaged ear is the one opposite the shoulder that the rifle rests on. The gun butt shields the ear that is next to it. In general, sustained noises like power boats and snowmobiles are more damaging than intermittent ones like rock music. Noisy working environments are especially dangerous. Routine diagnostic methods show no damage caused by loud music, but sophisticated laboratory tests reveal two abnormalities: less sensitivity to sounds at certain frequencies and less ability to separate complex sounds into component parts.

▶57 **What is the normal body temperature of a healthy human being? *Hint:* It's *not* 98.6°F.**

A. 98.2.
B. 98.5.
C. 98.9.

A, 98.2 overall. In 1868, Carl Wunderlich reported that he had taken more than one million measurements of body temperature on 25,000 patients. Using an under-the-armpit thermometer that took 15 or 20 minutes to warm up, he concluded that the normal temperature for most humans is 98.6°F. That figure has been used ever since.

Using modern digital thermometers in 1992, researchers produced a more accurate average normal temperature: 98.2°F or 36.8° C. Normal oral temperatures of adults ranged as high as 98.9°F in the early morning and 99.9°F overall. Over the course of a day, normal oral temperatures were found to vary up to 4.8°F between individuals and up to 1.09°F in a particular individual. Women are about 0.3°F warmer than men, and African-Americans are about 0.1°F warmer than whites.

▶58 During famines and wars, hungry Europeans called it "mountain meal." Starving Finns mixed it with flour and tree bark. Sardinians added it to acorn bread, while Indians in the Andes and southwestern United States added it to potatoes. Pregnant women everywhere ate it, too. What is it? And why is it eaten in so many places?

Clay, because it detoxifies poisons. Thanks to clay, people and animals can eat a wide variety of marginal and toxic foods and escape stomach upsets. Because clays are composed of fine particles with large surface areas, they bind easily with organic compounds. Moreover, the atoms in the clays are organized in layered sheets, making it easy for their atoms to change places with charged particles in poisons like the glycoalkaloids in bitter potatoes. With acorns, clay acts as a catalyst to alter the structure of tannins so that they cannot bind to proteins in the gut. As a result, pregnant women eat clay to calm their stomachs. Incidentally, humans are not the only creatures with this habit. Chimpanzees and gorillas, which eat many toxic plants, add clay to their diets. And rats, when poisoned, chew it too. Clay eating is such an established practice that it even has a proper name: geophagy.

▶59 How long did the average patient stay in a hospital in the nineteenth century?

A. 30 days.
B. 60 days.
C. 6 months.
D. 1 year.
E. 3 to 5 years.

B, D, E. 52 days in maternity wards; 1 year in general wards; and 3 to 5 years in "incurable" wards, according to records of the Philadelphia Almshouse and the Pennsylvania Hospital. In 1991, the average patient stay was only 5.6 days.

Throughout the nineteenth century, middle- and upper-class patients were cared for at home; only the poor went to hospitals. Because the very ill died relatively quickly, most hospital patients were not especially sick. They were given rest, warmth, and nourishing food rather than active therapeutic care. On a typical day at the Philadelphia General Hospital, "most of the beds are now unoccupied, because many of the patients are convalescent and are out in the yard sitting, smoking, or reading."

▶60 **If a lamb does not spend the first six hours of its life with its mother, the ewe rejects the child. But if a farmer stimulates the mother's genitals for five minutes, she nuzzles and nurtures her baby normally. What is the half aphrodisiac–half childbirth hormone that provokes such a response?**

Oxytocin, sometimes called the hormone of happiness and satisfaction. This small peptide released by the pituitary gland at the base of the brain may be the reason mothers and fathers cuddle each other and their children. The amount of hormone in the body surges during orgasm and induces a feeling of satisfaction afterward. During childbirth, it stimulates the mother's uterine contractions and makes her milk flow. In some species, it may even convince the father to build a nest and guard his offspring.

▶61 **Lawrence L. Craven, a general practitioner in Glendale, California, during the 1940s and early 1950s, noticed that many gum-chewing tonsillectomy patients developed bleeding complications. As a result of this observation, he began recommending a daily treatment for his middle-aged, male patients, especially those who were overweight and sedentary or who were recovering from a heart attack. His studies, based on 8,000 patients, were ignored for 40 years. What was Craven prescribing?**

Aspirin. His young tonsillectomy patients had been chewing aspirin-gum to relieve their pain. Of Craven's 8,000 male patients, not one had suffered a detectable case of coronary thrombosis or a major stroke. Today, it is believed that an aspirin a day keeps the doctor away—at least where heart attacks and strokes are concerned among middle-aged men.

►62 **The first cornea transplant was performed in 1905, the first blood transfusion in 1918, the first kidney transplant in 1954, and the first heart transplant in 1967. Besides organs, what can people donate for use after their death?**

The accompanying diagram indicates the full range of human tissues that can currently be donated. Bone grafts are the most common transplant operation; approximately 350,000 were performed in 1990.

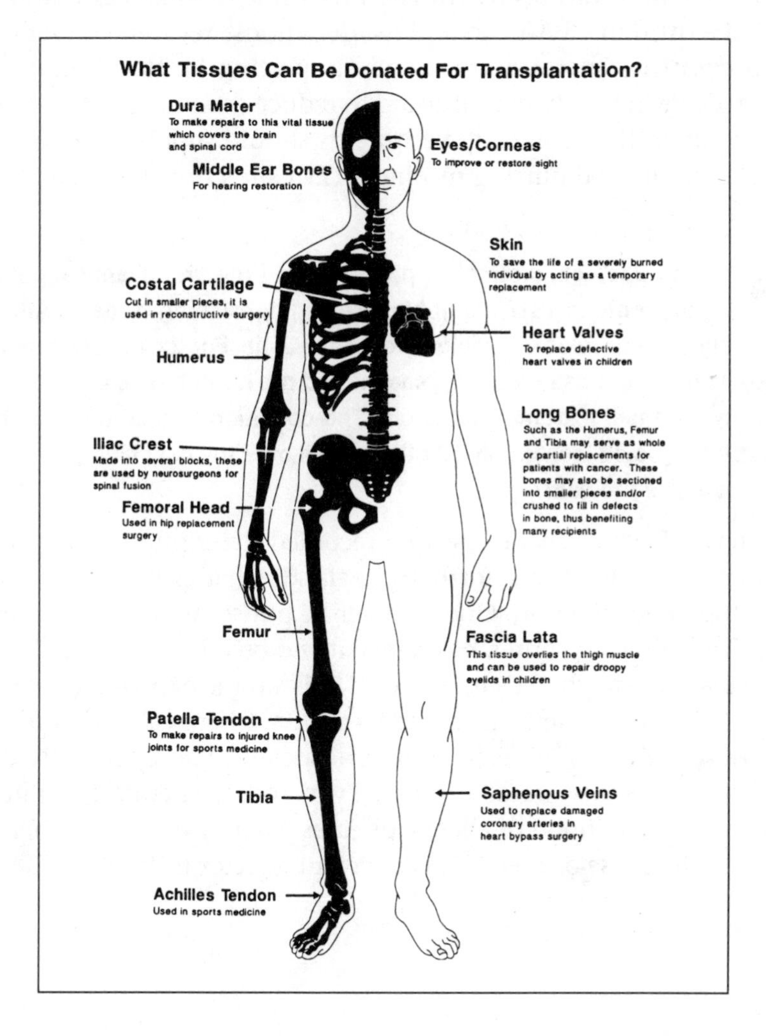

▶63 **What is the primary method of birth control for all nonseasonally breeding animals, including people?**

Lactation. Breast-fed babies have fewer diarrheal and respiratory infections and fewer allergies. But nursing is also good for mothers, especially in third world countries where contraception may be expensive, unavailable, or culturally unacceptable. In fact, breast-feeding averts more births in the third world than modern methods of contraception. A women who nurses her child on demand for six months and is not menstruating has the same 2 percent chance of conceiving as a woman who uses a modern method of reversible birth control. It has been estimated that nursing reduces fertility by almost a third in 18 African and Asian nations. Yet some underdeveloped countries spend more money importing bottle formula than they receive in family planning aid to reduce their birth rates. Instead of buying bottle formula, governments should supplement the diets of undernourished nursing mothers, *The Lancet* urged in an editorial.

▶64 **John Bostock, an English physician and the first chemical pathologist, wrote in 1817 about his rare and debilitating disease. After the age of eight, he developed a violent cold each June. For the next two months, he would blow his nose profusely, sneeze and rub his itchy eyes, breathe with difficulty, and feel extremely lethargic. The condition was so unusual, however, that he could locate only 28 other cases over the next nine years. What was Bostock's problem?**

Hay fever. Bostock made the first recorded description of hay fever. Another physician from northern England, Charles Blackley, suffered from the same affliction. One summer he put some summer pollen in a jar. The following winter, he opened it and breathed in the pollen. He immediately "caught" a summer cold. His simple experiment was elegant proof that hay fever is caused by pollen. Chemical pollution from the Industrial Revolution in northern England had apparently damaged these doctors' nasal mucosa. Even today, chemical pollutants increase the likelihood of allergic diseases like hay fever. Whether pollution is the cause or merely an aggravating factor is debated.

▶65 **One hundred million women are missing and presumed dead. Where and why?**

In Asia, India, and many other third world countries where boys are preferred to girls. Families abort daughters, kill them at birth, or simply give them less food and medical attention than boys. Western countries have approximately 105 females for every 100 males. In 1991, however, India had only 92.9 females for every 100 males; in 1991 China had only 93.8 females per 100 males. Afghanistan, Bangladesh, Bhutan, Nepal, Pakistan, Papua New Guinea, Turkey, and Egypt had fewer women than expected too. The number of missing women included 30 million in China and 60 million throughout Asia.

▶66 **These three people are attempting to relieve the same affliction: autointoxication. Autointoxication was the scourge of the late nineteenth and early twentieth centuries. It was blamed for 90 percent of all diseases, including baldness and premature senility. At its peak in the 1920s, it was "the universal menace," "the failure of civilization," "the foe of beauty," and the reason why men failed at business and women could not find romance and marriage. What was autointoxication?**

Constipation. Scientists and public alike regarded the contents of the bowel as the most ungodly filth of all. How could anyone be healthy, carrying around such a "cesspit" of filth inside them? As the maker of Dr. Pierce's Pleasant Pellets advised in 1895, "if fecal matters are retained until they are decomposed, great injury follows, since the fluid portions are absorbed, conveyed into the blood, and, of necessity, corrupt it with their impurities. . . . Sometimes the blood becomes so charged with fecal matter that its odor can be detected in the breath of the subject." As late as the 1930s, a team manager told his athletes that food is "a burden, fermenting, decomposing, putrefying, filling the body with poisonous substances, which are taken up by the blood stream, and this sewer-like blood flows all over the body."

The solution? Inner cleanliness. And how to achieve it? That was obvious: Use bran, purges, laxatives, cathartics, enemas, exercises, elaborate gadgets, and even surgery to remove or shortcut the large bowel.

Autointoxication became a veritable mania during the 1920s and gradually subsided thereafter. We, of course, are too sophisticated and knowledgeable for such nonsense. Even so, each year we spend $400 million on laxatives and make 2.5 million doctor's visits—all because of constipation. Bran muffin, anyone?

▶67 Baldness may or may not be bad for romantic love, but it definitely seems to be bad for the heart. That is, assuming it's one particular kind of baldness. What type is associated with an increased risk of heart attack?

So-called male pattern baldness, which starts as a spot on the top of the head in men under 55 years of age. Men who only have receding hairlines are unaffected. The mechanism linking baldness to heart attacks is still unknown. Other physical characteristics that are related to heart attacks are xanthomas (cholesterol-laden deposits close to the skin surface) and waistlines that are bigger around than the hips. Approximately one-third of middle-aged men have some degree of baldness.

▶68 Is vitamin D a vitamin or a hormone? And is the childhood disease rickets, which prevents proper bone development, caused by vitamin deficiency or sunlight deficiency?

It depends on whether you're talking to a nutritionist or a physiologist. Nutritionists regard cholecalciferol (vitamin D) as a vitamin, but increasing numbers of physiologists call it a hormone. Vitamins are generally considered chemicals that animals need but cannot synthesize themselves. Given enough sunlight, however, mammals can produce more than enough cholecalciferol. The ultraviolet component of sunlight converts 7-dehydrocholesterol in the epidermal skin cells to cholecalciferol. Any excess is stored in the liver until needed. Only when the body does not get enough sunshine are dietary supplements of vitamin D required. Once in the body, cholecalciferol functions like a hormone. In fact, the body converts it to a physiologically active hormone (calcifetriol), one of three hormones that together control the deposition of calcium in the skeleton.

▶69 **Pellagra is a vitamin-deficiency disease of poor people whose food staple is corn. Meat is a rich source of niacin but corn is not. A lack of niacin can cause skin, gastrointestinal, and mucosal problems as well as psychosis and damage to the central nervous system. Pellagra became a problem in Europe only when corn became a staple foodstuff in the Mediterranean during the eighteenth century. The disease was widespread in the southern United States well into the twentieth century. Yet pellagra was uncommon in Mexico, where corn tortillas are as important as bread. Why was Mexico spared pellagra?**

Traditional tortillas are made from corn flour treated with lime water. Normally, the niacin in cereals is so tightly bound to other chemicals that digestion cannot liberate it for use by the body. But when niacin is heated in the alkaline environment of lime water, it is liberated and can be digested.

▶70 **Speaking of cold-blooded and bloodthirsty murder, what animal holds the record for having killed the most humans?**

Mosquitoes. As disease vectors, they are the worst. Malaria alone has killed hundreds of millions of people; about 100 million new cases occur each year. Mosquitoes also transmit more than 100 viral diseases to humans and other animals, among them dengue, encephalitis, yellow fever, and filariasis.

▶71 **Medical schools in the United States were scandalously backward throughout the nineteenth century. What was the typical entrance requirement for the 155 medical schools in the United States?**

A. Elementary school.
B. High school diploma.
C. Two years of college.
D. College degree.

A, elementary school. Almost anyone could attend medical school in the United States. Almost anyone white and male, that is. Many accepted any male who had completed elementary school; only 50 medical schools required a high school diploma, or its equivalent. And only 16 of America's medical schools required two years of college. Johns Hopkins Medical School, founded in 1893, was the first medical school in the United States to require a college degree.

U.S. medical schools were money-making businesses owned and operated by local doctors. To compete for students, the schools kept their standards low and their courses short. Terms were often only eight weeks long. Instruction was by lecture only because laboratories were expensive. Canadian schools, modeled on those in Edinburgh, Scotland, were far superior. And medicine in Europe had already become a science.

▶72 **What do mesquite pods, prickly pear cactus, tumbleweed shoots, live oak acorns, amaranth seeds, and tepary beans have in common—besides their Southwest desert origin?**

They are traditional Indian foods of the Southwest that normalize blood sugar and suppress between-meal hunger. After World War II, the Pima and Tohono O'odham Indians abandoned their desert diet for high-fat city food. Today almost half of them develop diabetes by the age of 35, a rate 15 times worse than that of the general American population. Their traditional tribal diet of low-fat, high-fiber foods formed edible gels, gums, and mucilages. It also contained a slowly digested starch called amylose. Absorbed slowly over four or six hours, the foods prevented rapid swings in blood sugar.

▶73 **What do the following characteristics have in common?**

- **A combination of large molar, small incisor, and canine teeth.**
- **An adolescent growth spurt in both height and weight.**
- **A reproductive system involving permanent breasts, concealed ovulation, and continuous sexual receptivity in both sexes.**

Only humans have them. These traits set the human species apart from virtually all others. The combination of teeth is unique, and no other species—not even primates—have a growth spurt equal to that of adolescent humans. Adolescence begins with puberty and ends when adult height is reached, at about age 17 in girls and 21 in boys. The sudden and rapid growth rate of adolescents is faster than at any age except early infancy. Most other mammals grow at a constant rate.

▶74 **What specific part of a girl's skeleton continues to grow even after she has attained almost all of her adult height at age 16?**

The pelvic inlet, the bony opening of the birth canal. It develops more slowly than the rest of a girl's body and does not reach maturity until she is 17 or 18 years old. The slow growth of the pelvic inlet is the main reason why the vast majority of human cultures delays the birth of a woman's first child until she is 19 years old.

Worldwide, the average girl begins her adolescent growth spurt at about age 10; it peaks at about age 12. Approximately one year after the growth rate peaks, the typical girl begins menstruating. At this point, girls in most human cultures become engaged or even married. Most cannot bear children at that time, though, because most girls do not begin ovulation until they have been menstruating for one to three years.

Even after the average female starts ovulating, however, she is not reproductively mature. Her babies are more apt to have lower than normal birthweights for two major reasons. First, the girl's body is still growing taller, and her body gets first crack at her nutrients; the fetus gets the leftovers, so to speak. Second, the birth canal of the average female does not complete its growth until the mother is about 18. Only at that point is it big enough for a normal-weight baby to be born. Even cultures in which girls marry when they start menstruating generally delay the girl's transfer to her husband until she is reproductively mature at about age 18.

▶75 The average Chinese eats 20 percent more calories and is 20 percent *thinner* than the average American. True or false?

True. The Chinese eat twice as much starch and only a third as much fat as Americans. On average, the Chinese eat 25 percent less protein and get only 7 percent of the protein from animals; yet they suffer far less from heart disease and certain cancers. Moreover, although the average Chinese eats half as much calcium as a typical American, the Chinese rarely gets osteoporosis. And although the Chinese eat twice the iron but get it from plants instead of animals, anemia is no more common among adults than it is in the United States. Cancers of the reproductive tract and breast are rare in China. And the average life expectancy in China is 70 years, compared to 75 in the United States.

A study of 6,500 Chinese and their eating habits suggested that, to reduce the risk of heart disease and cancer, humans should eat only 10 or 15 percent of their calories from fat.

▶76 All the __________ in adults are formed in the fetus. None is formed after birth. After several years of activity, their structure changes dramatically. They stop growing, shorten, rest for a few weeks, and then begin development anew. What are they? (Fill in the blank.)

Hair follicles. Each of the 100,000 hairs on a human scalp is in a different stage of growth. A hair can grow three or more years before its follicle withers. After resting for several weeks, the follicle begins to produce a new hair, and the old hair is shed. On average, approximately 13 percent of the follicles on a human scalp are in the resting phase at any one time.

The follicle cycle is an evolutionary holdover from an annual cycle of molting. The synchronized growth and shedding of hair on seasonal mammals creates summer and winter coats with different insulation and camouflage.

▶77 An intestinal hormone functions in the human brain. What is the hormone, and what is it doing in two such different locations?

Cholecystokinin. In the small intestine, it helps digest carbohydrates, proteins, and especially fats. It is the chemical messenger that travels through the bloodstream to instigate the digestive process.

In the brain, cholecystokinin is a chemical messenger that transmits information across a synapse from one neuron to another neuron.

Thus, cholecystokinin exemplifies nature's economical use of one chemical to do two very different things. Since the discovery of cholecystokinin's dual role, other gastrointestinal hormones have been found to operate as neural transmitters too.

▶78 The brain cells of humans and other primates are irreplaceable; once damaged, they do not regenerate. So how can a canary's autumn song help an injured brain?

Unlike people, adult canaries replace old brain cells with newly made neurons. Thanks perhaps to those new cells, male canaries learn a new song each year. Their brain cell clusters associated with vocalizing actually grow larger during the spring, when males compose fresh melodies and females learn to recognize them. In the fall, the brain clusters shrink and the males forget their spring repertoires.

The growth of new neurons may help birds, which are quite long-lived, to acquire new information without having to carry around large and heavy brains. Perhaps human beings do not replace their old neurons because to do so would destroy their memories. In the future, however, it may become possible to induce canarylike neurogenesis in adult human brains.

▶79 A graduate student in linguistics hit her head in an automobile accident. Afterward, she became more assertive with her husband. From a country where wives defer to men, they were both shocked. For example, she had always wanted to see the movie that he recommended. Now she wanted to choose her own film. Dismayed, the young man filed for a divorce. What had caused his wife's scandalous behavior?

Postconcussion syndrome. Even a mild head injury can cause a year of headaches, fatigue, and alcohol intolerance. Intellectual abilities may suffer, including memory, creativity, motivation to do mental activities, and the ability to deal with more than one type of information at a time. And, like the young wife, victims may undergo personality changes and exhibit noticeably less insight and tact.

Blows to the head can damage the white matter of the brain. The white matter is made of axons, which, like wires, carry information from one part of the brain to another.

▶80 **Are these spheres convex or concave? Turn the page upside down and look at them again. Now which group is convex and which concave?**

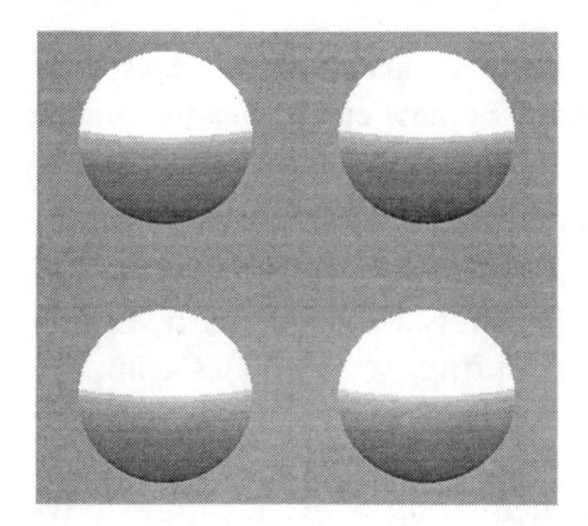

At first, the spheres in the group on the left appear convex, whereas those on the right appear concave. But when the pictures are turned upside down, as in the group on the right, the depth of the spheres reverses. Those on the left now appear concave, whereas those on the right appear convex. The brain assumes that the light comes from above. After all, our brains developed in a world in which the sun shines above.

But perhaps your objective knowledge of up and down has influenced your decision. To find out, turn your head upside down and look at the page. It is the orientation of the pattern on the retina that matters.

Conclusion: The eye forms two-dimensional patterns of light and color on a single plane of cells in the retina. Yet thanks to shading, the brain makes us perceive solidity and depth. The brain simplifies its job by assuming a single light source.

▶81 **A 17-year-old girl in Washington, D.C., suffered a deep leg wound, and a blood infection, meningococcemia, caused clotting in the leg.**

Oral antibiotics were ineffective. To control the sores and help them heal, physicians at Children's Hospital resorted in desperation to an old battle-ground remedy. They applied 1,500 maggots, or fly larvae, to the wound. The maggots ate away the dead skin, keeping the wound clean and enabling the healthy skin to thrive. Maggots are not the only unusual repair methods being used today, however. For example, today deep wounds are packed with a substance that Egyptian military surgeons used 4,000 years ago. It promotes new tissue growth by drying out the wound and dehydrating bacteria. It's cheap, easy to use, and painless. What is it?

Sugar. German physicians, in particular, pack deep wounds with sugar, although some prefer honey.

▶82 **They're used today after microsurgery in which fingers, toes, and other body parts are reattached. They prevent tiny blood vessels from clogging. What are they?**

Leeches. The vampire annelid worm has an anticoagulant in its saliva to keep the blood of its victim flowing. Its anticoagulant, hirudin, has been isolated and used—sans leech—to treat hemorrhoids, rheumatism, thrombosis, and contusions.

In terms of sheer numbers used, leeches must be one of the most popular treatments in history. In 1833, France alone imported 42 million leeches.

▶83 **Some researchers say, "Don't transplant organs. Use substitutes instead." So they implant cells on biodegradable polymer sponges placed in rats. As the cells grow, they take over some of the functions normally performed by an organ in the rat. Name the organ.**

The liver. Liver cells deposited on a polymer sponge are fed by capillary veins that grow into the sponge. As the liver cells multiply, the body breaks the sponge down into water, carbon dioxide, and other components, which are excreted in the urine.

▶84 **Eighteenth-century Eskimos made rain parkas from walrus and seal intestines because the translucent material was water impermeable. Today, the food industry uses pig intestine for sausage casings, and cooks**

use it for chitlins. Now pig intestines may find a medical use as _____. (Fill in the blank.)

A universal tissue graft. The diaphanous inner layer of pig intestine may become the template for the human body to replace worn-out blood vessels, ligaments, and bladders. The inner layer of pig intestine is made of collagen, the connective tissue that holds most organs together. After transplantation into animals of several different species, it disappeared completely. It had been replaced by the type of tissue it had replaced: whether blood vessel tissue, ligaments, tendons, or bladder muscle.

▶85 **To correct major congenital abnormalities, to bridge fractures with a graft, or to replace jaws destroyed by cancer, try _______________. It does not alarm the body's immune system or cause inflammation. And bone grows right into it and makes an almost seamless meld. What is this wonder material? (Fill in the blank.)**

Coral. The numerous interconnected channels in coral give it an almost identical physical configuration to bone. Before it is used, the coral is heated to kill the organisms in it and to convert its calcium carbonate to hydroxyapatite (an important constituent of bone). Only the coral's porous mineral structure remains. In time, blood vessels and bone spicules fill the channels and permeate it completely.

Ecology and Animal Behavior

▸86 **For decades, students have been taught that viceroy butterflies discourage predators by looking like bad-tasting monarch butterflies. So what's wrong with that?**

Nobody bothered to taste a viceroy. They taste awful too, according to a group of Florida red-winged blackbirds tested in 1992. The birds, participating in a laboratory experiment, became noticeably agitated after tasting the viceroys and refused to eat anymore.

So if the two butterflies taste alike, why do they look alike too? Presumably because a joint advertising campaign reinforces their common message to predators.

▸87 **Domesticating animals is apparently a tricky proposition because only 16 species have been tamed successfully so far. How many of these were domesticated by our neolithic ancestors?**

12, two-thirds of all domesticated species. Dogs, sheep, goats, pigs, and cattle were tamed between 10,000 and 15,000 years ago. About 5,000 years later our ancestors became tired of carrying their own burdens and domesticated donkeys, horses, two types of camels, llamas and alpacas, and buffaloes. The cat was their last conquest. The Romans, despite their great interest in wild animals and menageries, managed to tame only one species: the rabbit. Modern society has added only three more: hamsters, guinea pigs, and laboratory rats.

▶88 **Name a mammalian mother so masculinized that she:**

- **mates and bears young through an erectile clitoris the size and shape of a penis.**
- **has a pseudoscrotum in place of a vulva.**
- **produces testosterone within her placenta to bathe her babies in male hormones.**

The spotted hyena of sub-Saharan Africa. Newborns of both sexes produce their own male hormones. They are so aggressive that they begin fighting each other even before they are out of their amniotic sacs. Siblings of the same sex fight to the death.

As adults, spotted hyenas are so ferocious that a group of 22 can strip a 300-pound (135 kilogram) antelope to bare bones in 13 minutes and then crunch up the bones like so many potato chips.

Other hyena species are solitary scavengers. Spotted hyenas, however, are highly social animals, live in groups, and compete for food. An abundance of testosterone—linked to aggression in many species—helps them compete. But it also produces masculinized anatomy and infighting.

Spotted hyenas and domestic pigs are the only newborn mammals known to fight, but eagles and large fish-eating birds like herons, egrets, and boobies also indulge in brutal sibling rivalry.

▶89 **Only 12,000 years ago, western North America was a New World Serengeti filled with large mammals and birds: bison, camels, antelopes, mammoths, saber-toothed cats, giant dire wolves, tapirs, condors, and so on. One or two thousand years later, roughly 73 percent of the large mammal genera from the late Pleistocene era were extinct. Most of the big bird species had died out too. Why?**

Probably because of people, specifically an invasion of Paleo-Indian hunters of the Clovis culture. Skilled big-game hunters, they would have been stalking animals that had no prior experience with human predators. The same phenomenon happened when human beings arrived in New Zealand, Madagascar, and Australia: Their sheltered "megafauna" disappeared soon after their first contacts with human hunters.

Paleo-Indian hunters arrived in North America as the climate was warming in North America, Europe, and Asia. Yet climate did not wipe out the European and Asian "megafauna"; the big animals were extinguished only in North America. Those in Europe and Asia were accustomed to human predators and were more wary, although they too were eliminated eventually.

▶90 **They are different species, but buddies all the same. Hunting together, one supplies its keen smell, its hearing, and its burrowing prowess. The other is the pouncer. Who are these hunting pals, whom the White Mountain Apaches called "Cousins That Travel Together"?**

A badger and a coyote, who hunt ground squirrels as a team. The badger locates the prey and digs into its tunnels. The coyote guards the escape routes and catches the fleeing squirrels. Badgers and coyotes have hunted together during the prime summer months for centuries. A pre-Columbian pot pictures a badger on one side and a coyote on the other. And Aztecs regarded the badger as the "coyote of the earth."

▶91 **They'd make a soap opera couple: the archetypal femme fatale paired with the classic, self-sacrificing male. While the female cruises coolly from one group of males to another looking them over for a possible mate, the male whips himself into a frenzy searching for his once-in-a-lifetime love. After mating, he collapses from the excitement and dies. The female lives on, tightly controlling the sex of her young to avoid competition from daughters. Her babies are among the smallest in the mammalian world. Who is this soap-opera family?**

Meet the antechinuses. Called the Australian "marsupial mouse," an antechinus is a tiny, shrewlike tree-nester. It weighs only $\frac{1}{448}$ of an ounce ($\frac{1}{16}$ of a gram) when born and only about $1\frac{3}{4}$ ounces when adult. Despite its tiny size, the antechinus has become the darling of biologists because of its complicated sex life. Once-in-a-lifetime mating, mass postmating death, and strict control over their offspring's sex are all rare among mammals. Litters of one species of antechinus sometimes exhibit the most extreme male bias ever recorded among mammals, whereas litters of another species are almost equally skewed toward females.

Most biologists today believe that species develop particular lifestyles to maximize the number of their descendants. Antechinuses face intense competition for survival. They are the second most common species in the east coast of Australia.

▶92 **What is the most extensive habitat on Earth—and one of the richest, as well?**

The surface of the world's oceans. This 3-foot-thick habitat covers more than 71 percent of the planet. Thousands of species of plants, animals, and microbes call it home.

Its surface microlayer—the topmost, paper-thin layer roughly two-thousandths of an inch thick—is host to a dense concentration of minerals, chemicals, protozoans, and microorganisms. The next few inches down contain larger organisms—fish eggs, fish larvae, and crustaceans— that feed on the microorganisms above. Within the upper foot live larger floating jellyfish and seaweeds. Scientifically, the region is called the "sea surface layer." Its unique ability to collect and condense chemicals makes it especially vulnerable to pollution.

▶93 **What animal is all of the following?**

- **The world's only cold-blooded mammal.**
- **The only mammal that lives like a social insect, that is, like an ant, a honeybee, or a termite.**
- **A farmer that grows food crops, yet feeds its feces to its children.**
- **A vocalizer as skilled as some primates.**
- **One of the homeliest creatures known.**

The naked mole-rat *Heterocephalus glaber,* which is not naked and is neither mole nor rat. It lives in underground colonies of 70 to 80 animals in the rock-hard semidesert of Ethiopia, Kenya, and Somalia. Related to guinea pigs, chinchillas, and porcupines, a naked mole-rat looks like a saber-toothed sausage or a loose-skinned walrus.

A naked mole rat. This chubby chewer is widely considered to be one of the homeliest creatures in the animal kingdom.

Like a queen bee, a queen mole-rat bears her colony's young. To keep her slim enough for tunneling while pregnant, her vertebrae expand lengthwise. She lives with one to three mates and the young from previous litters, who maintain and defend the group like the workers in social insect colonies. By shoving and poking would-be breeding females, the queen keeps them under so much hormonal stress that they cannot come into estrus or mate.

Mole-rats eat tubers, and nonbreeders "grow" food by refilling half-eaten tubers with soil. Nonbreeders feed their feces to the queen's pups to provide them with both nourishment and the intestinal microorganisms for digesting cellulose.

As its fans boast, a mole-rat is "not just another pretty face." Mole-rats are chubby, chewing machines. One-quarter of their muscle-mass is concentrated in bulldoglike heads and jaws. In laboratories, they have chomped through concrete. In East Africa, a wild colony of 87 mole-rats dug 200 meters of tunnel a month, excavating 350 kilograms of soil from an area as large as 20 football fields.

With heavy inbreeding, colony members share an almost identical genetic makeup. Nonbreeders contribute to the survival of their genes by helping the entire family endure in a harsh environment. To biologists, the homeliness of the mole-rat is only skin deep.

▶94 Heather Kakapo, regarded as one of the world's weirdest and rarest birds, delighted conservationists in 1990. What did Heather do?

Heather laid an egg.

Kakapos are giant parrots from New Zealand. They are also flightless and nocturnal. Like the extinct dodo, the kakapos led sheltered island lives until settlers arrived. Since kakapos nest on the ground and must leave their eggs and young to search for food, their nests were easy prey for marauding rats introduced by Maoris and by cats and stoats introduced by Europeans. By 1990, only 43 kakapos remained in the wild.

For safety's sake, Heather and 21 other parrots were moved to a cat-less island 1,100 kilometers (700 miles) away. When Heather produced her egg, conservationists took it as a sign that the move had not upset the flock unduly. By 1993, 44 kakapos were protected from alien mammals in "safe" island refuges. There they are fed supplemental rations to encourage yearly breeding.

▶95 How many species of feather mites make their home on a green conure parrot in Mexico?

Approximately 30. And since each mite goes through four different stages on its way to maturity, the plumage of one green conure can house 120 highly particular kinds of creatures. Conures are veritable flying zoos. As such, they have come to symbolize the complex interdependence of highly specialized species, especially those in tropical rain forests.

Feather mites eat the oily secretions and cellular detritus of avian plumage. They are highly specialized. A particular species may specialize in a particular site on a particular bird's particular feather, for example, the outer quill of a primary wing feather. Seven species of mites have been counted just on one feather.

The eighteenth-century satirist Jonathan Swift had a different animal but the same principle in mind when he joked:

So, Nat'ralists observe, a Flea
Hath smaller Fleas that on him prey;
And these have smaller fleas to bite 'em
And so proceed *ad infinitum*.

▶96 Who was the first person bitten by a bedbug?

Probably a caveman—or woman. Two-thirds of the 89 species of bedbugs are blood-drinking parasites of bats. Only three species of the family Cimicidae bite people. Those three probably evolved to take advantage of cave dwellers who shared their quarters. Bedbugs require a meal of blood if their eggs and nymphs are to mature.

▶97 **Why are some male birds of paradise glamorous and sexually promiscuous, while others are drab and faithful to their families?**

You are what you eat. In species that eat a nutritious, mixed diet of complex fruits and insects, a female easily feeds her young without help from a male. In those species, the males tend to be gorgeously plumed and promiscuous. They have plenty of free time for elaborate displays and competitions for females.

In species that eat exclusively fruit, however, females need help foraging for their young. Often mated for life, the males have no need, or time, for fancy plumage.

▶98 **Food sharing, even in emergencies, is extremely rare among mammals. Yet the females of one species form a long-term sisterhood, sharing food and caring for one another's young. What is the animal?**

The common vampire bat *Desmodus rotundus,* which must consume 50 percent to 100 percent of its body weight in blood each night. *Desmodus rotundus* dies of starvation if it misses a meal two nights in a row. Yet despite the precariousness of their existence, vampire bats often live 18 years, and females form long-term relationships for 12 or more years. How do they manage it?

Two females form a buddy system, sharing their blood meals when one of them has gone without food for 24 hours. They even share food with their partner's offspring. And most remarkable, the buddy does not even have to be a relative.

▶99 **How many different life-forms exist on Earth?**

A. 100,000.
B. 1,000,000.
C. More.

C, somewhere between 2 million to 30 million. The wild disparity between these two estimates shows that we do not have even a minimal understanding of life on Earth yet. Thousands of new species are discovered yearly, including three species of monkey during the early 1990s.

▶100 **Tropical rain forests, which occupy 6 percent of Earth's land surface, are home to __________ of the species of organisms. (Fill in the blank.)**

A. More than half.

B. A third.

A, more than half. As a rule, the number of species increases as one moves from the poles to the equator. For example, only about 20 species of land birds breed in Alaska in an area that would have 600 species in Central America. Only a few types of trees grow in Canadian, Siberian, and Scandinavian forests while tropical rain forests host thousands. Most of the world's undiscovered species live in the tropics, where a stable climate ensures a constant food supply even for highly specialized niche species.

▶101 **The Amazon Basin holds the record for hosting the most species of native birds. What percentage of the world's bird species spends time there?**

Thirty percent of the world's 9,040 bird species. Another 16 percent live in Indonesia. Most of these areas are rain forest or associated riverine and swamp woodlands.

▶102 **Whether they're cow pies or meadow muffins, buffalo chips or dung pats, the average elephant produces approximately 50 pounds of them a day. So why aren't we up to our eyeballs in elephant dung?**

Because African dung beetles, known more elegantly as scarabs, come to the rescue. Africa alone has more than 2,000 species of dung beetles. With heads like bulldozer blades, dung beetles dig in to dine. In their case, waste makes haste, and competition to grab and store the tasty morsels is fierce. At sundown, clouds of beetles swoop down on fresh elephant dung, burying football-sized boluses in minutes. Some species tunnel under the droppings and truck tidbits to their lairs. Others mold dung into balls and roll them away for burial. Still other species coat the balls with clay; early settlers to Africa and India confused the orbs with cannonballs. After burying the dung, the beetles—or the larvae they lay in the balls—can dine at leisure underground.

Egyptians venerated scarab beetles because their larvae, encased in dung, emerge from it with new life. According to one theory, the pyramids are symbolic dung pats enclosing the pharaohs until their rebirth.

▶103 **Using tools, sharing food, and hunting cooperatively are three characteristics that separate human beings from other animals, some theories say. What animal belies this long-held truism?**

Chimpanzees in Täi National Park in the tropical rain forest of the Republic of the Ivory Coast. Chimpanzees there collect and save stones of different sizes to crack open hard-shelled nuts, skills they learn as children from their mothers. Hunting as a group, chimpanzees circle their prey, coordinating their movements with one another *and* with the victim. Then the chimps share the spoils of the hunt with one another and their children.

The sophisticated behavior of these rain forest chimps also argues against theories that the development of humans was spurred by dry climates in the savannahs of eastern Africa.

▶104 **When wasps threaten to chop a metalmark caterpillar into bits to feed their babies, the caterpillar summons an army of ants to scatter the foe. How does the caterpillar summon troops from another species?**

With a cheery little song: 23 pulses of sound per second at an average frequency of 1877 hertz. The butterfly caterpillar sings its tune by rubbing specialized appendages together.

In return for their help, the ants feed on amino acids that are dispensed by the caterpillars near their tails. More than 100 tropical butterfly caterpillars are now known to communicate with their ant defenders by singing. All are members of the *Riodinidae* and *Lycaenidae* butterfly families. Their intensely symbiotic relationship is typical of many others in tropical rain forests, where the stable climate allows extremely complex biological communities to develop.

▶105 **Birds congregate by the tens of thousands on island rookeries to raise their young. But where and what is the world's largest concentration of mammals?**

A. A bat cave outside San Antonio, Texas.

B. Pribilof Island during seal mating season.
C. Reindeer herds in the Arctic.
D. The Milwaukee Zoo.
E. The Serengeti.
F. Lemmings in northwest Norway.

A. Twenty million female bats and 20 million bat pups summer in Bracken Cave. Adult Mexican free-tailed bats (*Tadarida brasiliensis*) mate in Mexico during the late winter. Then the females fly north at altitudes of 8,000 to 10,000 feet at an average speed of 40 miles per hour. Only about 100,000 males accompany them.

Using a vaginal secretion to keep their mates' sperm in a state of suspended animation, the females transport the sperm north. Only when they reach their creche caves in the southern United States is the sperm implanted so that pregnancy can begin. Returning to Bracken Cave from her evening food hunts, the mother needs only a few minutes to identify her pup from among 20 million others. She remembers roughly where she left her offspring in the cave and then narrows the field by smell and call.

▶106 Within a decade, a population of 30 million bats that wintered in a cave outside Phoenix, Arizona, was almost destroyed. What killed them?

Too many visitors and researchers. Bats are extremely shy, and their enormous cave populations are fragile. Each time a hibernating bat was disturbed, it used up several days' worth of its food supply; arousing them more than a few times each winter killed them.

Bat populations are worth saving, if only for their usefulness to people. Twenty million nursing bats consume 150 tons of insects nightly within a 60-mile radius of the cave. Bat mothers must eat half their weight in insects every night. Chewing and digesting their prey on the fly, so to speak, they take three-hour nursing and rest breaks in the middle of each eight-hour foraging trip.

Although these bat populations seem spectacularly large, mammals in general are outclassed as groupies by birds. Red-billed queleas roost in the tens of *millions* in Africa. And the extinct passenger pigeon formed two-*billion*-bird roosts covering several thousand acres.

►107 **What is the fastest disappearing group of vertebrates?**

Bats. The only mammals that fly, bats account for nearly one in four of the known species of mammals. Yet 40 percent of the world's 1,000 bat species are listed as endangered or threatened. Texas has more bat species than any other state in the United States.

Insect-eating bats emit high-frequency sounds and locate prey by their echo; some bats can sense an echo delay of 69 to 98 *millionths* of a second or detect an insect as thin as a human hair. Large bat colonies typically devour a half million pounds of insects nightly. Fruit-eating bats are responsible for the pollination and seed dispersal for important crops like bananas, avocados, vanilla beans, peaches, and figs.

The bat's only serious natural enemy is people.

►108 **As any reader of Agatha Christie mysteries knows, a quiet English village can be a murderous place indeed. What role does the pet cat play in real-life bloody dramas?**

The villain: Killer Cat. British cats catch an estimated 70 million creatures yearly—64 percent of them small mammals and 35 percent of them birds. And that's assuming the darlings bring home all their victims. A study in the United States indicates that cats bring back only about half their victims—er, that is, half the *number* of victims they kill. In which case, cats in bird-loving Britain may kill at least 40 million birds yearly.

The record catch is held by a feline from Dorset that brought its proud owner 400 victims per year. Mice, voles, and shrews accounted for most of the small mammals caught while 16 percent of the catch consisted of sparrows. Perhaps one-third to one-half of all sparrow deaths can be attributed to pet cats.

Now, *there's* a murder plot for you!

►109 **Bluefin tuna are the biggest bony fishes in the world and among the strongest and fastest. Capable of short sprints at 50 miles per hour, they can cross an entire ocean 5,000 miles wide in 50 days. Who or what is their worst predator?**

Japanese sushi lovers. A fully grown bluefin—at least eight years old and 310 pounds or more—is highly prized for its oil-rich meat in Japan, where it can fetch $10,000 to $15,000. Between 1970 and

1990, the western Atlantic population of fully grown bluefins plummeted 90 percent from 319,000 to 30,000. Only giant bluefins are capable of spawning.

In 1991, bluefin tuna were the first commercial fish proposed for inclusion on the international list of endangered species.

▶110 Wood and varying lemmings always produce more daughters than sons. On average, they bear three or four females for every one male. How?

Many of the females have a Super-X chromosome that overwhelms the male-producing Y-chromosome. When the Y-chromosome is unable to express itself, a daughter is born. Lemming populations undergo boom and bust cycles. During a bad year, few survive; in these small, isolated, and inbred communities brothers may wind up mating with their sisters. Hence, a female with a Super-X chromosome who produces more daughters than sons contributes more offspring to the next generation.

Presumably, a lemming with an aberrant but powerful version of its X chromosome passed the trait to its offspring and gave them such an advantage that the Super-X chromosome eventually spread throughout the lemming population.

▶111 Plump 'possum mothers produce primarily sons: 1.4 sons for every daughter. Weak and scrawny females produce predominantly daughters: 1.8 daughters for every son. Why?

The female opossum adjusts the proportion of male and female children in her litters, according to the availability of food. Well-fed mothers can produce sons that are strong enough to compete with other males for mates and to impregnate many females. But when food is scarce, an ill-fed mother can produce more descendants by concentrating on daughters. Each daughter is likely to produce a litter, while weak sons cannot compete for mates.

Opossums confirm a theory that when female mammals have the resources to lavish care on offspring, they produce more males than females. The females, so the theory goes, will find partners and reproduce even when they are weak and hungry. Weak males, on the other hand, may end up without mates or offspring, and so they benefit most by whatever extra care their mothers can give them.

▶112 **For gracious living, it's hard to beat the jewel wasp. Smaller than a fruit fly, this parasitic wasp lays its eggs in the blowfly pupae that inhabit dead carcasses and bird nests. If the jewel wasp female is the only one laying eggs at a particular site, 85 percent of her children will be daughters. But when several jewel wasps nest nearby, half her offspring will be males and half females. Why is she so gender-biased?**

A solitary jewel wasp produces enough males to inseminate all her daughters but not enough to cause wasteful competition among the sons. In a group, she needs enough sons to compete successfully for mates against the males of other jewel wasps.

You get the picture: Under some conditions, mothers are better off having sons. In other environments, females are better off with daughters. And some of the time, she can exert some control over the kind of children she will bear.

▶113 **Red-cockaded woodpeckers live in the few remaining stands of mature pine trees in the southern United States. Why do they fledge almost two sons for every daughter?**

Nesting sites are extremely scarce, because of habitat destruction. So young males help out their parents by incubating, guarding, and feeding later litters. The woodpecker is one of the few bird species in North America in which nonbreeding individuals cooperate to pass on the genes of the family as a whole.

▶114 **Some of the oldest communities in the Pacific Northwest date back long before the Mayflower landed at Plymouth Rock. Their leaders, females of tarnished reputations, live to be 70 or 80 years old. They and their descendants remain together their entire lives, so long that each group has developed its own dialect. Who are they?**

Killer whales, the largest members of the dolphin family. The killer whale population of British Columbia lives in matrilineal groups of 10 or 20 related animals. They are among the few animals that can reproduce the sounds they hear. Only humans, birds, a few primates, a few seals, and the dolphin family have that skill. And outside of birds and humans, local dialects are even rarer.

Zoology and Animal Physiology

▶115 **Plunging deeper than any mammal known, this animal dives for *months* nonstop day and night. Occasional breathers for oxygen take only two or three minutes. One female rested only 2.6 minutes after 62 minutes underwater. Another was tracked more than 4,125 feet under the surface of the ocean, four-fifths of a mile deep. A third dove up and down like a piston for 34 consecutive days and nights. What are these superlative divers?**

Female northern elephant seals (*Mirounga angustirostris*). While nursing its pup on land for a month, the elephant seal fasts from food and water and loses 40 percent of her normal, 650-pound weight. Then she copulates, weans the pup, and jumps back into the water. For the next two and a half months, she dives continuously up and down like a yo-yo. She forages in the twilight zone off the continental shelf, far deeper than other marine animals and commercial fishing go. There she can feed without competition on hake and squid. After another month on land to molt, she returns to the water.

▶116 **What is the secret of the elephant seal's spectacular performance?**

Elephant seals use oxygen stored in their blood, not in their lungs. They have enormous quantities of oxygen-rich blood in their bodies: two and a half times what a human of comparable size would have. Their blood is also extra rich in red blood cells, the oxygen carriers of the body. In fact, during most of their dives, elephant seals do not use their lungs at all. They *ex*hale before going down and hold their breath. At 130 feet, their lungs collapse and stop functioning. Their hearts slow down, and blood circulates only to vital organs and skeletal muscles needed for swimming. By using only the oxygen stored in blood, they avoid the lactic acid buildup that gives humans sore muscles.

Amazingly, elephant seals use less energy diving than swimming or resting on the surface. With their metabolism reduced by an estimated 40 or 50 percent, they dive in a torpor state, sleeping while diving. How they withstand the rapid changes in pressure is still unknown.

►117 The hoatzin bird is "a flying cow," and a poorly flying one at that. But it might as well be called "a flying brewery." How did the hoatzin get its bovine nickname?

It has the digestive system of a cow or sheep. Only a few bird species subsist on leaves, and the hoatzin is the only nonmammal that ferments leaf cellulose in the foregut, like a cow. Up close, the bird reeks of Eau de Toilette, otherwise known as cow dung.

Plant-eating animals use microbes in the gut to ferment cellulose into simple, usable sugars. But such a fermentation vat takes up so much space that few mammals or birds under 11 pounds can specialize in leaf-eating. The hoatzin, however, weighs only about a pound and a half. How does it manage to both eat and fly?

To digest its leaves, the hoatzin has an enormous crop and lower esophagus. They account for 10 percent of its weight. The crop is so large that the bird's chest pad is calloused from leaning against a branch after meals. Like leaf-eating sloths and koalas, the hoatzin has a low metabolic rate and spends most of its days lazing around, waiting for the crop to do its job. Food enters and exits most birds in minutes, but the hoatzin dawdles 24 to 48 hours over solids, like a sheep.

In case you wondered: Ptarmigan, grouse, and ostriches also eat leaves, but their fermentation vats are horselike, in their hindguts.

▶118 **Busy as a bee, an eager beaver, the little red hen, and the industrious ant—such are the fables we feed our children. Have we been leading them astray?**

Yes. While biologists spend much of their time studying animal feeding, fighting, and fornicating, many animals spend most of *their* time resting. They laze around—to conserve energy, save calories, digest meals, cool down, warm up, or stay camouflaged.

Lions, for example, spend 75 percent of their time resting, digesting huge meals. Moose, which are ruminants, need four hours to metabolize their food. Even ants and bees may work only 20 percent of the time. Beavers work only about five hours a day, and sloths, of course, are legendary: They sleep 15 hours a day and move so seldom that two algae species inhabit their fur and claws. Hummingbirds are the world's most energy-intensive fliers, but they spend 80 percent of their days and 100 percent of their nights perched on twigs. It makes no sense to race after food unless the odds of catching something are quite high.

Such R&R seems downright amateurish, however, compared to the spectacularly long rest periods that some organisms spend in dormancy (plants and insects); in diapause (the state of reduced metabolic activity seen in many insects); and in hibernation (frogs, snakes, salamanders, and many mammals).

▶119 **No animal, not even a cheetah, can beat the pronghorn American antelope for long-distance running. The pronghorn can comfortably cover 10 miles in 15 minutes and average 45 miles per hour. Most mammals can maintain peak anaerobic activity for only a few minutes, because it produces lactic acid as a waste product. How does the pronghorn avoid doing the same?**

Pronghorns can process three times more oxygen than other animals their size. Their lungs are spectacularly large, three times those of comparably sized goats. Their hearts are larger too, and their blood is richer in hemoglobin, the pigment that delivers oxygen to the muscles. Finally, their muscle cells are more densely packed with mitochondria, the structures that burn oxygen for fuel.

Thus, as an oxygen user, the pronghorn is more like a mouse than a man. Small mammals consume oxygen faster than large animals.

One gram of shrew muscle consumes as much oxygen in one *day* as a gram of elephant tissue does in a *month*.

▶120 What color is a polar bear's fur?

Colorless and transparent. Polar bears are solar bears; they have fiber-optic fur. Each hair is a hollow fiber tube that operates like the fiber optics used for telephone and computer communication. In fiber optics, modulated information moves through multilayered transparent fibers on beams of light.

In the polar bear, rays of ultraviolet light are trapped at the opening of the tube. From there, they are reflected off the inside surface of the tube down to the bear's skin. Polar bear skin, which is black, absorbs the ultraviolet light as heat. A bear can absorb 25 percent of its energy requirements from the sun in summertime, saving energy to hunt food and to accumulate fat for winter. Though other arctic mammals also have fur that absorbs ultraviolet radiation, the polar bear is tops, with an efficiency rating of 95 percent.

Polar bear fur looks white because of the way the rough inner surfaces of the hair tubes reflect visible light. Incidentally, by the end of the feeding season, the fur often looks yellow. Polar bears are not neat eaters. The yellow is stain from seal blubber oil.

▶121 Which is harder on polar bears: warming up or cooling down?

Cooling down. With thick fur, hide, and blubber, polar bears rarely have to burn extra energy to keep warm. In fact, in infrared photographs, resting or slow-moving polar bears don't even show up. The only heat detectable is a tiny patch in front of the nose—exhaled breath. Thus, as long as it stays out of the wind and is relatively still, a polar bear does not have to change its metabolic rate, even in temperatures as low as −37°C (−34°F).

On the other hand, they can't exercise heavily without heating up. At −20°C (−4°F), polar bears cannot walk faster than 4 kilometers (2.5 miles) per hour without raising their temperatures. Lumbering along at only 7 kilometers (4.5 miles) per hour, their temperature would rise to a feverish 37.8°C (100°F).

As a result, polar bears don't stalk prey. They lie down, motionless and sometimes asleep—and wait for seals to come to their air holes. That way, the seal uses its energy to get caught, not the bear's. Cool work, if you can get it.

▶122 What mammal mates in the spring, but does not become pregnant until the fall? *Hint:* Its babies are the smallest of any placental mammal in proportion to their parent's size. Raised on superrich milk, its young have one of the fastest, or fattest, growth rates on record.

The polar bear. Like the European badger and many other carnivores, female polar bears mate and their ovum is fertilized, but its implantation is delayed. Thus, gestation begins in the fall, and young are born in midwinter. A baby polar bear grows from the size of a rat to the size of a human in one year.

▶123 "Our near relatives in the animal world are by and large an extraordinarily murderous lot," according to one biologist. In some primate groups, the leading cause of infant death is ___________. (Fill in the blank.)

Infanticide. More than a dozen species of primates practice infanticide, including New and Old World monkeys and apes. When the male leader of a group of animals is ousted, the new head male often kills the infants sired by his predecessor. That way, the females in the group come into estrus sooner and the lead male does not have to wait until their babies are weaned to mate with the mothers.

A study of infanticide in Canada showed that a human baby is 70 times more likely to be killed if he lives in a family that includes a stepparent instead of two parents. Almost always, the murderer is the male stepparent.

▶124 Musk, an aroma produced by a gland in Tibetan deer, is a staple of the perfume industry. Why are people so fond of its aroma?

Perhaps because the steroid musk molecule is very similar in shape to the male sex hormone testosterone. Women are a thousand times more sensitive to the smell of steroid-type molecules than men. So who do women wear perfume for?

▶125 What bird cruises 10,000 miles in search of food for a month at a time at speeds up to 50 miles an hour, all the while leaving its mate back at the nest meal-less and hatching their egg?

The wandering albatross *Diomedea exulans*. It flies the strongest and most regular winds in the world, those of the southern oceans between Antarctica and South America, Africa, and Asia. Its foraging trips in search of squid take the bird far beyond the range of land-based radio receivers. Consequently, little was known about their travels until the advent of satellite tracking. The largest marine bird, a wandering albatross weighs in at 25 pounds and stretches 11 feet from wing tip to wing tip. A tendon between their wingbones locks their long, slim wings open so they can glide for hours—or days—without muscle power. In fact, a wandering albatross requires only a third more energy to fly than to sit on its nest. It has the most efficient flight of any marine bird. Only the most efficient gliding aircraft outperforms a wandering albatross. The bird never flies straight upwind. Although it heads into the wind, it tacks back and forth for days without flapping its wings.

▶126 Working, nursing mothers may envy the hooded seal. After only four days of nursing, her baby is weaned and on its own. No other mammal known has such a short lactation period. How does she do it, and why?

Hooded seal babies (*Crystophora cristata*) are born on unstable Arctic ice that is ready to break apart, float away, or melt at any moment. Growing up triple-time is advantageous. Precocious even before birth, the babies begin fattening up in utero. When born, they already have 1.5 centimeters of blubber. Then, in a world where "fat is fortune," they slurp the world's richest milk: 61 percent fat, creamy enough to make anyone but a seal babe gag. In four days of nursing every 40 minutes around the clock, the mother transfers a third of her fat reserves and almost a fifth of her total weight to her offspring. Drinking 17 to 20 pounds of cream a day, the pup gains about 15 pounds of weight daily. Then, her work over, the mother migrates to the open seas for food. The baby rests—stuffed, sleepy, and contented—untended for a week. By late April, when Arctic ocean food supplies are at their richest, the youngster is ready to swim, eat, and survive, on its own.

▶127 **The acorn woodpecker of California is a champion jackhammer. Its beak bangs into wood at 20 to 25 kilometers (12 to 15 miles) an hour. The force on its head is a thousand times the force of gravity and 250 times what an astronaut experiences during takeoff. How does the woodpecker survive without brain damage?**

Its braincase is made of spongy bone joined with opposing muscles that work like shock absorbers. In addition, its head jabs back and forth in one plane as accurately as a metronome. If it rotated, its brain would be jostled loose and destroyed.

▶128 **If you were a bacterium in the ocean, how would you orient yourself in the dark?**

You would use your compass, naturally. Many kinds of bacteria make their own magnets. Using iron and either oxygen or sulfur, they form iron mineral crystals, each about 50 nanometers or 50 billionths of a meter long. The crystals line up along the North to South magnetic pole, making a tiny compass needle in each cell of the bacterium. Then Earth's magnetic field orients the compass needles and their cells as the organism swims through the water.

Bacteria in the Northern Hemisphere head toward the North Pole; bacteria in the Southern Hemisphere swim toward the South Pole. In the Northern Hemisphere, Earth's magnetic field points downward as well as northward. So bacteria following Earth's magnetic field are directed downward as they swim. This enables them to reach the ocean's food-rich sediments. In the same way, bacteria in the Southern Hemisphere reach sedimentary levels by swimming southward.

Natural magnets can account for up to 2 percent of the bacterium's total weight. Biomineralization, the production of minerals by living organisms, was once thought to occur only in higher organisms.

▶129 **If life 800 miles from the North Pole weren't so hazardous, arctic woolly-bear caterpillars would live 14 years. Of all the dangers that they face, what is the greatest?**

A. Parasites.

B. Winter temperatures that can dip to −68°C (−90°F).

C. Summers.

A and C. Arctic woolly bears (*Gynaephora groenlandica*) are supercool caterpillars. They spend most of their lives frozen in a state of suspended animation. They come to life for only a few weeks each summer to feed and grow. Then, they go back into cold storage to avoid the pesky parasites that flourish at the height of the Arctic summer. Parasites kill five times more woolly bears than winter does, making summer the most dangerous time of year for these cool critters. Winter's cold kills only about 13 percent of them. With so little time to grow, Arctic woolly bears need 14 years to mature.

▶130 If bumblebees are busy in temperate climates, they are even busier in the Arctic. To jerry-build a colony fast during the short polar summer, the queen bee produces fewer but bigger litters. She steals her nests ready-made from lemmings and snow buntings. And she and her fellow bees shiver to warm up their flight muscles to 30°C (86°F) or more. But not even this fast-paced lifestyle enables the bumblebee queen to complete her colony before winter's cold begins. So what does she do?

She starts incubating her eggs even *before* she lays them. And to do it, she raises the temperature of her abdomen as if it were an oven. No bird or mammal known can independently regulate the temperature of its body parts in response to its environment. But then, the queen *Bomus polaris* is one of only two of the world's 20,000 species of bees to live above the Arctic Circle.

Incidentally, when news of the incubating bumblebee was announced, the *National Enquirer* scandal sheet published an "exposé," lambasting scientists for studying "the rectal temperature of a bumblebee."

▶131 What 250-pound glass-eater becomes the featured, but sometimes fatal, attraction at feasts?

The hawksbill turtle, *Eretmochelys imbricata,* the second most endangered species of marine turtle. The hawksbill dines on sponges found in tropical coral reefs. The sponges are composed of a hydrated, amorphous silica similar to opal, which is a form of glass. As a result, the turtles' gastrointestinal tracts are chock-full of glass splinters and their feces are virtually solid glass.

Occasionally, South Pacific islanders who dine on hawksbills at ceremonial banquets die within 24 hours, probably because of blood-destroying compounds and metabolic blockers in the sponges. The hawksbill is endangered, not because of its diet, but because its beautiful shell is used for jewelry and for Japanese wedding combs.

▶132 Sabrina holds the record for feline pesematology in New York City. Who is Sabrina, and what did she do?

In 1987, Sabrina the Cat fell 32 stories onto a New York City sidewalk and survived with only one chipped tooth and mild chest injuries. Feline pesematology is the science of falling cats from the Greek word for "fall," *pesema*.

Records show that cats survive long falls even better than short ones. In fact, while 90 percent of cats survive falls from two to six stories, 95 percent survive falls from seven to thirty-two stories. Cats have a superb gyroscope in their inner ears. Within the first 2 or 3 feet of fall, they right themselves so that all four legs point down. They land with their limbs flexed, distributing the force over four legs.

When they fall more than seven stories, other factors also help. At terminal velocity, they no longer feel the pull of acceleration. Apparently as a result, they spread their limbs out like flying squirrels. This increases their air resistance, decreases their velocity, and spreads the landing's impact over their body.

Researchers now believe that nine may be a conservative estimate for the number of lives enjoyed by urban-dwelling felines.

▶133 Consider the following facts about the hooded pitohui bird of New Guinea:

A. Both males and females have the same brilliant black and orange coloring. Insects mimic the colors.

B. Their New Guinean name means "rubbish."

C. According to the traditional theory, frogs and insects produce poisonous toxins to protect themselves against predators, but birds do not.

What's your conclusion about the theory?

A hooded pitohui bird.

The theorists in C are wrong. Birds can produce poison. The hooded pitohui (of the genus *Pitohui*) is the first documented case of a poisonous bird. It uses one of the most lethal poisons ever discovered: the toxin used by Amazonian hunters on the tips of their blow-darts. The hunters get the chemical from a South American frog.

Several pieces of experimental evidence should have warned biologists that their theory in clue C might be wrong: They knew that bright plumage among birds is associated, inversely, with tastiness. A series of banquets featuring 200 bird species during the 1940s and 1950s demonstrated that the blander the bird looks, the better it tastes. And the fact that males and females share the same colored plumage should have warned biologists that the male's feathers were not designed to attract females.

It was serendipity, not pure reason, that led to the discovery, however. In 1992, a graduate student licked his hands after handling the bird and felt his mouth burn and go numb. Biologists were so surprised that the *New York Times* put the story on page one.

Incidentally, a diet of poisonous insects or berries may be responsible for the bird's fatal chemistry. Now biologists suspect that more of the world's 9,200 bird species use chemical defenses too. The problem is: Who wants to taste them?

▶134 Human runners have improved their racing times almost five times faster than thoroughbred horses. What's the problem with the horses?

Horses breathe in and out with each galloping stride. When they run faster, they take longer strides but do not breathe more often. Thus, they cannot increase their oxygen intake. Nor can they eliminate lactic acid fast enough as it builds up in their muscles. Most important, thoroughbreds have not developed the skeletal strength to support great increases in speed. Bone fractures are common among outstanding horses. Great racehorses may be athletes pushed to the verge of destruction.

▶135 Mammals give birth to live young; birds lay eggs; and amphibians deposit their eggs in water. Right?

Wrong. Or at least not always right. More than 60 species of tree frogs in the rain forests and mountains of Central America and tropical

South America are marsupials. That is, their eggs develop on the mother's back, often in a marsupial-like pouch. The babies emerge as small, fully formed frogs or as precocious tadpoles.

Tropical forests are host to so many frog species that egg sites are hard to find. Mother's back is one of the few locations not already spoken for.

▸136 **"What's different?" asks Irene Pepperburg. "Shape," responds Alex. Why has Alex, the African gray parrot, been named "Honorary Flying Primate"?**

Because Alex *Psittacus erithacus* proves that parrots can be as intelligent as primates. A student at Northwestern University and the University of Arizona, the parrot has learned to categorize the words for 80 objects by color, shape, and material. Alex can also identify quantities up to six. Even more surprising, he understands abstractions like relative size or the absence of things. Emotionally, Alex is the equivalent of a two-and-a-half or three-year-old child.

In the wild, parrots are intensely social, live in large flocks, and form permanent relationships. When alone in captivity, they easily become bored and depressed. Scientists advise anyone who keeps a large parrot as a pet—as distinct from a domesticated budgerigar—to spend as much time interacting with it as with a toddler. Because large parrots live 50 years or more, arrangements must also be made in case they outlive their owners.

Better yet, don't keep any large parrot as a pet. Of 330 parrot species, an estimated 77 are in immediate danger of extinction, thanks to habitat destruction, trapping and hunting, and a low reproduction rate.

▶137 **What North American animal is for its size among the shortest-lived mammals in the world?**

Opossums. In the wild, most live only one season. The few that survive into a second year show obvious signs of aging, including cataracts, weight loss, and poor motor control. One of the few wild mammals known to show physiological signs of aging, opossums are now used in gerontological research.

▶138 **What opens its mouth and swallows its prey faster than any other predatory vertebrate? Within six milliseconds, it expands its mouth 12-fold to suck in its victim. Virtually any prey entering the strike zone of this voracious creature is dead meat.**

Hint: **It changes color to match its surroundings, uses bait to lure its victims, walks and jet-propels, and is considered one of nature's most advanced examples of aggressive mimicry.**

A frogfish. It attracts prey by wiggling elements of its elongated dorsal fin to resemble bait at the end of a lure. Depending on the species, the frogfish can change color within seconds to match its background. And it can sneak up on prey by swimming, by hobbling about on leglike pectoral fins, or by jet propulsion (ingesting enor-

Is it a rock encrusted with algae or a warty frogfish?

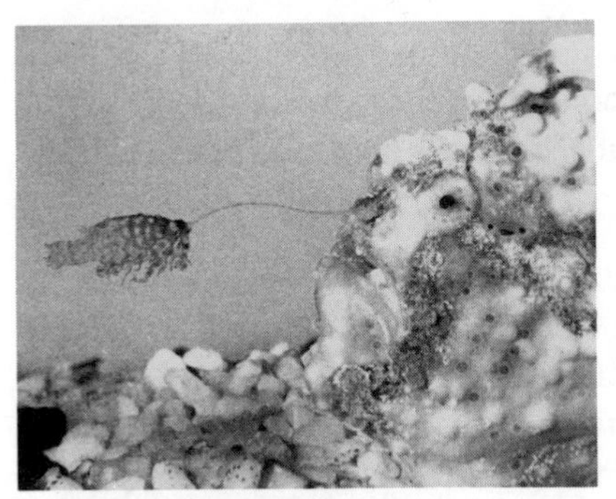

Is that a delectable fish swimming by or a lure enticing prey to an untimely death?

When another creatures happens by, the frogfish wiggles its lure.

mous amounts of water and then forcing it backward through its gills like a jet). No other fish has as many highly evolved adaptations. Frogfish live in tropical and subtropical waters around the world.

▶139 What is the only warm-blooded mammal known to survive, without freezing, body temperatures below freezing?

The supercooling, hibernating arctic ground squirrel, *Spermophilus parryii*. When its body temperature drops as low as −2.9°C (26.8°F), it uses one-tenth the energy it would need to maintain its temperature above zero.

The animal apparently uses some method of supercooling so that its fluids cannot freeze at or below their freezing point. Its blood does not contain antifreeze molecules, however. The arctic ground squirrel is the first mammal discovered that can survive after being supercooled for more than an hour.

▶140 What male animal has become such a specialist in "childcare" that *he*—and *only* he—gets pregnant?

The seahorse (*Hippocampus*). After the female deposits eggs in the male's pouch, the male fertilizes them. Then he supplies them with oxygen from his capillaries; releases the hormone prolactin to bathe the embryos in a nourishing "placental" liquid; and gradually changes the fluid to seawater to protect the newborns from shock. In some species, males and females form long-term pairs, mating again as soon as the male gives birth to their young.

▶141 Hair is a weapon if you're a porcupine with needlelike hairs or a rhino with a horn made of hair. What animal uses its hair as both a warning signal *and* a projectile?

A tarantula that is the world's largest spider: the mouse-eating spider (*Theraphosa leblondi*) of the Guiana region of South America. With a 10-inch leg span, it weighs in at almost a quarter of a pound.

When threatened, the tarantula rubs its hairy appendages together to produce a staticlike hissing that is audible for several meters. If a foolish predator persists, the tarantula continues its rub-down until a little cloud of hairs is released. Barbed and sharply pointed, they make a predator's skin burn and itch for days. Moths are the only other animals that can produce such a severe reaction with their hair.

Only as a last resort does an embattled tarantula bite.

A Theraphosa leblondi tarantula.

▶142 What warm-blooded creature can live to be 100 years old; travel 10,000 miles a year; distinguish between squares and diamond shapes and between red and white; detect electric fields equal to a 1.5 volt flashlight 2,000 miles away; zip along at 30 miles per hour, and do all this without ever sleeping?

Sharks, considered among the best-built organisms in the natural world. Thanks to a unique circulatory system, sharks are warm blooded and several degrees warmer than the ocean they inhabit. Blood warmed by muscle activity is transported to the heart by veins that encircle the arteries and warm their blood. Sharks lack the gas bladder that makes most fish buoyant, so they must literally sink or swim. Their livers also help keep them afloat. Shark livers produce oils that are lighter than water. Since a shark's liver can occupy up to 90 percent of its body cavity and contribute 25 percent of its weight, the organ actually decreases the overall density of the shark's body.

▶143 The skin of the nuclear submarine was pitted with craters the size of Vanilla Wafers. Can you identify the cookie cutter?

A shark species (*Isistius brasiliensis*) the size and shape of a police officer's billy club. It cuts round plugs into the skin of whales, sea lions, porpoises, tunas, and billfishes. Then it rotates its body as its suction-cup mouth sucks out the flesh. The shark also liked the looks of the submarine—and tried it out. Incidentally, the cookie cutter is one of the most luminous sharks. Hundreds of light-producing organs on its underside and head make the shark glow an eerie green in deep, dark water.

One-third of the world's 350 known shark species have been discovered since 1970. Sharks range in size from 6 inches to 45 feet. The two smallest shark species, between 6 and 8 inches long, are dwarf dogsharks (*Etmopterus carteri* and *Etmopterus perryi*) found off the Caribbean coast of Colombia. Sharks can be scavengers, parasites, plankton eaters, or bottom feeders. They live in deep and shallow ocean water, in freshwater, or on the sea bottom.

▶144 How many people are attacked worldwide by sharks each year?

Between 50 and 75 worldwide. Of these, fewer than 30 may be fatal. Most sharks are small and harmless, according to the International Shark Attack File in Gainesville, Florida. The three most dangerous sharks are the white shark, tiger shark, and the bull shark. The oceanic whitetip shark (sometimes confused with reed whitetip sharks) may also be dangerous.

People, on the other hand, kill millions of sharks a year. Sport and commercial fishermen are responsible. Commercial fishermen often catch sharks in their drift nets, kill them or cut off their fins for sale, and then throw the fish back into the ocean. Sharks reproduce slowly; they mature late and give birth to only a few young. So fishing is thought to have depleted 20 or more species.

▶145 What is the world's most primitive dinosaur, the great-granddaddy of them all, the creature that is just a few evolutionary steps from the common ancestor of all dinosaurs?

Eoraptor lunensis. Eoraptor, a dog-sized dinosaur, was named "Dawn Stealer" because it lived at the dawn of the dinosaur age about 225 million years ago and probably relied on stealth and speed to hunt. Half of the world's 350 known dinosaur species have been identified since the 1970s.

A nearly complete Eoraptor fossil was discovered in October 1991 in northwestern Argentina in a 50-mile-long rift region in the foothills of the Andes Mountains. About 25 pounds in weight, Eoraptor measured about a meter from nose to tail. It was a quick-moving meat-eater with hollow bones. It walked on hind legs and grasped its food with three-fingered claws on short arms high on its chest. Eoraptor apparently evolved before the branches in the dinosaur family tree separated. Its fossil supports theories that dinosaurs began as small, bipedal carnivores about 230 million years ago.

▶146 What was the first vertebrate?

The conodont, a 2-inch-long eel with a finned tail. Scanning electron microscope studies of the fish showed in 1992 that its gripping teeth were made of bone cells, enamel, and a supporting layer of mineralized cartilage. Bone, whether in the spinal column or elsewhere, characterizes vertebrates. The development of bone was an important evolutionary step leading toward higher animals and the human race.

Conodonts—the name means "cone teeth"—flourished 515 million years ago in shallow Cambrian seas, long before animals moved onto land. Conodonts became extinct 200 million years ago, during the heyday of the dinosaurs. The analysis of conodont bone came only a decade after the discovery of the creature's soft tissue in fossils. Only then was it known that the conodont had looked like an eel with a finned tail.

▶147 What mammal:

- **has a venom gland in its hind foot that can kill a dog?**
- **has extraordinarily large mammary glands a third its length?**
- **is a nocturnal bottom feeder in freshwater streams, where it can maintain its internal body temperature for several hours in subzero water?**
- **has retractable webbed feet, a bill like a duck, and a tail like a beaver? Its anatomy is more reptilian than mammalian; that is, it lays eggs (though it nurses the hatchlings) through the same cloacal opening used for excretion.**

The platypus (*Ornithorhynchus anatinus*). This eastern Australian native used to be considered a primitive creature ill-adapted to modern life. Recent studies show it to be both highly specialized and well adapted to life in freshwater streams and lagoons.

▶148 As the world's only egg-laying mammals, platypuses and echidnas have broken numerous world records. But recent news about them electrified biologists: The platypus and the short-snouted echidna of Australia are the only mammals that get a charge out of hunting. What's going on?

They are the only mammals known to detect weak electric fields. Many fish and amphibians receive or generate electrical discharges to locate or stun prey. Their electrosensory systems developed from sense organs that help them detect pressure waves in water. In Australia, the platypus and echnidna use a different technique: electroreceptors on their fleshy snouts. The duck-billed platypus uses them to find shrimp in cloudy rivers, while the echidna's snout locates beetle grubs underground.

The Australian echidna is one of the few dry-land animals with electroreceptors. Electric fields generated by water currents and animal muscles are extremely weak and are more easily detected in water than on land. But the Australian echidna lives in arid semi-deserts and rain forests. So it supplies its own water: its perpetually runny nose.

▶149 Many human parents would love to select the sex of their children—usually to have sons. Nonhuman species generally produce equal numbers of sons and daughters. But some species routinely produce lopsided sex ratios among their progeny. How do egg-laying reptiles control the sex of their young?

Not by chromosomes, but by the incubation temperature of their eggs. American alligators, for example, produce more males at warm temperatures while some turtles develop more males at cool temperatures.

▶150 Why does the normally discreet blue grouse display such gaudy garb while courting, and why does he waste so much energy showing it off? The obvious answer is, "To attract a female." But no biologists worth their salt would settle for such simplicity. What might they say instead?

Far from being a foolish waste of resources, the male's elegant garb and flashy behavior prove to the female that he is healthy and strong.

A blue grouse being discreet.

A blue grouse displaying his finery.

Long and exhausting songs, dances, and other rituals may show that the male is free of parasites and debilitating conditions. Females can distinguish among males on the basis of their physical characteristics. For example, a female gallinule (also called a moorhen) prefers her males sleek and fat, and she can detect the mate that fits the bill. Female barn swallows prefer males with long and symmetrical tails, while Japanese scorpion flies prefer males with symmetrical bodies. Why symmetry? It may be a sign of good genes. Or it may simply be style. So far, no one knows for sure.

►151 Who are the hotheads of the fish world?

The few bony fish—including tuna, swordfish, marlins, and butterfly mackerel—that have literally "warm-blooded" brains.

Swordfish and marlin pass their blood through a special eye muscle to warm their brains and eyes. Tuna conserve muscle-produced heat in their muscle, brain, and other vital areas before the heat disappears out their gills. Each of these hotheads lives in a wider range of water temperatures than their cold-blooded relatives. Tuna, for example, migrate from tropical to subpolar seas, and swordfish dive between the warm surface and the cold depths to catch squid. Only a few dozen of the 30,000 species of bony fish have developed techniques to warm vital parts of their bodies.

▶152 **A few male bullies control all the females and feeding territories among a particular species of fish in Lake Tanganyika. Each bully is big, fiercely macho, and emblazoned with orange stripes. The nonmating males of the species are small, mild-mannered, and sandy-colored. But a bully's appearance, aggression, and sexual prowess depend solely on social status. And that changes radically when another, even larger male appears on the scene. Obviously, the social position of the first bully plummets. Next, his orange stripes turn sandy, his testicles wither, and his bluster melts away.**

Has the loss in prestige destroyed him?

Almost. The male *Haplochromis burtoni* cichlid fish is a spectacular case of mind over matter, social environment over biology, and behavior over brain structure. When the bully experiences a loss in prestige, his brain cells actually shrink. These are the cells in his hypothalamus that enabled him to bully other fish and to mate. In top-ranking males, these neurons swell six or eight times bigger than those in low-ranking males. Even more intriguing, the male's status, and hence his flamboyant appearance and behavior, can be changed back and forth in the laboratory fish tank by introducing or removing a larger male.

Plant Science

153 What is the world's largest living organism?

A. The blue whale.
B. A giant sequoia.
C. An aspen grove.
D. A fungus.
E. Water weeds.

C. A 106-acre, 6,000-ton stand of genetically identical quaking aspen stems is one individual plant. Located in the Wasatch Mountains south of Salt Lake City, Utah, the entire grove of 47,000 aspen stems grows from one root system. Although the stems look like separate trees, they are actually just different parts of the same plant. They cloned naturally, forming a single individual. Every leaf in the grove is the same size and shape, and they change the same color at the same time each fall. Although individual stems may die, the organism as a whole can survive for long periods of time.

Traditionally, blue whales and sequoia trees have contended for the title of "Biggest Living Organism." But the largest blue whales weigh only about 100 tons. And the aspen clone weighs more than three times as much as the largest known giant sequoia.

Quaking aspens are the most widespread tree species in North America.

▶154 **What is the oldest organism known to inhabit the earth?**

A humongous fungus in northern Michigan spawned from a single spore 1,500 or more years ago. The fungus invades 30 acres underground and weighs an estimated 100 tons, as much as a blue whale. The giant *Armillaria bulbosa* fungus—known as the button or honey mushroom—spreads underground near Crystal Falls, on the Wisconsin border. It probably began growing at the end of the last Ice Age. The fungus, a web of mushrooms and rootlike tentacles, feeds on dead wood and other decaying matter. Detailed genetic analysis proved that the fungus is one discrete unit. The discovery raised a fundamental question: Just what is an individual unit in biology?

In 1992, when it was discovered that the Michigan fungus is one plant, scientists thought that it might be the world's largest organism too. But an even larger, 500- to 1,000-year-old fungus was found near Mt. Adams in the state of Washington. And an aspen grove in Utah proved to be bigger than either humongous fungus.

▶155 **Imagine living for thousands of years in frozen Antarctica high on a rocky, windswept mountain as dry as a desert. Where would you seek shelter?**

In a rock. Lichen living *inside* porous sandstone rocks are the only organisms known to grow high in the dry, rocky mountain ranges of Ross Desert, Antarctica. Temperatures there rarely rise above freezing, and hurricane-strength winds evaporate the moisture. But north-facing rocks that soak up polar sunlight can be 15°C (60°F) warmer than the surrounding air. And sandstone is porous enough to absorb melted snow. Lichens—symbiotic partnerships of algae and fungi—take advantage of this relatively cozy environment to form black, white, and green stripes a few millimeters below the surface of the rocks.

Cryptoendoliths, organisms that live "hidden in rock," are found in deserts worldwide. But Antarctica's lichens barely survive. As photosynthetic organisms, they use the sun's energy to convert carbon dioxide and water into food. But they use 99.9 percent of the area's available resources to stay alive; normally, plants save 10 percent of their resources for growth and reproduction. Barely holding their own, Antarctica's lichens grow exceedingly slowly. They are thousands of years old and may be among the oldest organisms on Earth.

▶156 A pox on exotic orchids, endangered species, and other rare specimens. What are the world's most common plants?

Two minuscule planktons that live in the world's oceans: prochlorophytes and a type of cyanobacteria. Each plant is tiny: one fiftieth the diameter of a human hair or less. But they are fabulously abundant almost everywhere they have been sought: from the Pacific to the Gulf of Mexico, the Caribbean, and the Atlantic Ocean.

The most common plant, identified in 1988, is a prochlorophyte. There are as many of these cells in 10 gallons of water as there are people on Earth. This rare phylum, discovered in the late 1970s, has only three members. Barely visible with traditional microscopic techniques, it was identified using a laser-based flow cytometer, a machine more commonly used in medical research.

The second most common plant form discovered so far is a type of marine cyanobacteria or blue-green algae found in 1978. It has only one type of chlorophyll, the pigment molecule that captures sunlight and converts it into organic matter. While there are 10^4 of its cells per millimeter, there are 10^5 prochlorophyte cells per milliliter.

▶157 A rose is a rose is a rose. But a mustard flower may not be a mustard flower. It may be a pseudoposy, a fraudulent flower, a fungal fake. If it is not a mustard blossom, what is it?

An imitation flower perpetrated by a rust fungus, one of the botanical world's greatest con artists. When the mustard plant *Arabis holboelli* is infected with the fungus, it produces a fake yellow "flower" even more beautiful, odiferous, and nectar-filled than its own normal blue blossom. The "flower" is actually a cluster of yellow leaves that looks like a buttercup to both insects and botanists. The leaves are yellow because they are coated with the male and female sex organs of the yellow fungus. Flies, bees, and butterflies faced with a field of flowers spend more time hanging around the frauds than the gen-

The modest, uninfected mustard plant Arabis holboellii.

The mustard plant infected by a rust fungus grows a fake flower.

A butterfly feeds on sugary fluid in the pseudo-posy.

uine articles. Insects thus spread the gametes of the rust fungus, instead of those of the diseased mustard plants.

Incidentally, there are about 5,000 species of rust fungus, many of which infect only one particular species. In this case, the mustard plant was infected with *Puccinia monoica* or *Puccinia thlaspeos.*

▶158 Why should lovers perplexed by the opposite sex ponder the lowly slime mold?

Because the slime mold (*Physarum polycephalum*) doesn't have two sexes to worry about; it has 13 sexes. Humans can exchange chromosomes with only half the people on earth: those of the opposite sex. But slime molds come in 13 different genetic types, so any slime mold can choose a mate from among 12 other types, or sexes. Can you imagine a Hollywood production, *Tales of the Sexy Slime Mold*?

▶159 Will the real First Flower please step forward? Leading contenders for the crown are magnolias, black pepper plants, water lilies, and hornworts. The correct answer will help explain the evolution and structure of plants *and* their pollinating animals. Birds, insects, and bats evolved along with flowers and, by spreading their pollen and seeds, created natural hybrids between closely related plant species. All of today's flowering plants evolved from one ancestral line; they share too many common features to have

descended from several different plants. But which one is the real "First Flower"?

Paleoherbs, specifically water lilies and black pepper plants, seem to be the current odds-on favorites. They have tiny, simple, and unisexual flowers. Another possibility, however, is the hornwort, an aquatic plant with neither petals nor sepals. It is found in freshwater ponds and lakes but resembles no other living flowering plant.

Until recently, the choice would have been the magnolia, which combines both male and female sex parts in one large blossom.

Whichever it was, flowering plants brought color to green earth and increased the degree of specialization among both flowers and their pollinating animals.

▶160 For the biggest caffeine jolt of all, what plant should you brew?

A. Coffee.
B. Tea.
C. Cacao.
D. Kola nuts.
E. South American holly.

E, the South American holly, *Ilex guayusa*. It contains the highest percentage of caffeine by dry weight of any plant in the world. In Ecuador and Peru, macho males of a remote Amazonian tribe drink tea made from its leaves; one cup has as much caffeine as five cups of coffee. To avoid the jitters, sweats, and headache that characterize a caffeine overdose, the men immediately vomit up most of the brew. Maté, a more familiar herbal tea, is made from another South American holly, the *Ilex paraguayensis*.

▶161 Sand trap, speed trap, claptrap, What makes the plant world's fastest trap?

The Venus flytrap *Dionaea muscipula* Ellis. The jawlike leaves of the carnivorous Venus flytrap need only one to three seconds to curl shut on an insect. This is an almost nervelike response. Except for its relative *Aldrovanda*, the Venus flytrap has the fastest cell expansion in the plant world.

The key to the Venus flytrap's fast action is rapid growth, which, in turn, is probably caused by a rapid production of acid. When an insect brushes against trigger hairs on the leaves of a Venus flytrap, electrical pulses activate a powerful cellular pump. The pump removes hydrogen ions from the leaf cells, creating an acid on the outer surface of the leaf. The acid, in turn, stimulates the cells on one side of the leaf to expand and close the leaf. The leaves use up almost a third of their energy closing their traps.

▶162 The world's biggest botanical pest has one of the highest biomass production rates anywhere. In just *one* growing season, 25 plants can generate two *million* plants that cover 10,000 square meters and weigh as much as a loaded jumbo jet. Name this beautiful troublemaker.

Water hyacinth, *Eichhornia crassipes*. It is the world's most aggressive aquatic weed. Thanks to its air-filled tissue, it floats from one body of water to another. And because its rosettes break off easily, it can quickly clone into huge mats up to 40 kilometers (25 miles) long and 2 meters (6 feet) deep. The Amazonian native clogs the waterways of more than 50 tropical and subtropical countries on five continents.

▶163 Believe it or not, this plant can double its biomass in 2.2 days. Reproducing clonally, the world's entire population of this plant is genetically the same individual organism. If all its specimens formed one continuous plant, it would be the largest individual organism on Earth. What is it?

The kariba weed, whose scientific name *Salvinia molesta* is a clue to its pestiferous personality. A Brazilian native, it grows five times faster than the water hyacinth. Kariba forms meter-thick mats that clog waterways in Africa, Asia, and Australia. Thanks to a Brazilian beetle, however, kariba is not the world's most troublesome plant pest. The beetle feeds exclusively on kariba plants. So thanks to its finicky appetite, the kariba plant is being brought under control. Introduced into Australia and Papua New Guinea, the beetle has munched its way through hundreds of square kilometers of the weed. Within two years, kariba beetles had gobbled up two million tons of the plant. Then they started nibbling their way across India and Namibia.

▶164 As meticulously as a surgeon, a beetle severs the midrib vein of a leaf. When sticky liquid oozes from the wound, the insect moves on to nibble the tip of the leaf. What's going on?

Both the red milkweed beetle and the caterpillar of the monarch butterfly operate on the veins of milkweed leaves before eating them. White poisonous latex is stored under pressure inside canals alongside the veins in the leaves. When the veins (and canals) are severed, the noxious latex oozes out. After "milking" the canals, the insects feed freely on the down-vein end of the leaf.

Many caterpillars, beetles, and katydids perform similar premeal surgery on plants that store latex, resin, and other unpleasant secretions in canals. If damaging one vein does not eliminate the offending liquid, an insect may cut a long incision all the way across the leaf.

▶165 One of the world's major food crops is so poisonous that it is grated to a pulp, soaked, fermented, squeezed, and roasted before it is eaten. Name the food and its poison.

Cassava (*Manihot esculenta*) and cyanide. Cassava, eaten in virtually all tropical countries, grows in poor, dry, and acidic soils. It resists pest damage and disease and is one of the most efficient converters of solar energy into carbohydrate. Tapioca is composed of rice-sized beads of cassava starch. Despite its value as a Third World food, cassava is being replaced by subsidized grains grown in the Northern Hemisphere.

▶166 What plant takes longest to produce a seed?

Bamboo. Most bamboo species flower and produce seeds only once in a lifetime, when plants are between 12 and 120 years old. A forest of bamboo flowers simultaneously and dies soon afterward, leaving a blanket of seeds 20 to 30 inches deep. Fortunately, bamboo also reproduces asexually by sending up underground shoots that are genetic clones of itself. Unfortunately, these clonal rhizomes cannot be used for breeding new, improved hybrids.

After years of experimentation, scientists succeeded in 1990 in making bamboo flower in a laboratory. Their achievement should pave the way for the bamboo's hybridization and reforestation. The secret of their success? Feeding tiny cuttings of bamboo on coconut milk, growth hormones, and other nutrients.

▶167 Fred had his Ginger, and Romeo had his Juliet. Why does a Madagascar orchid need a giant moth with a 15-inch tongue?

The rare Madagascar orchid (*Angracecum longicalcar*) has a 16-inch-long hanging tube that collects nectar at the bottom. The tube's lining is constructed to block crawling insects, so the orchid can be pollinated only by a flying creature with a body part that can be inserted into the tube.

When Charles Darwin studied a related Madagascan orchid with a tube only a foot long in 1862, he predicted that a moth with an 11-inch tongue must pollinate the flower. In a virtuoso confirmation of Darwin's theory of evolution by natural selection, the moth was discovered and described in 1903 by the Rothschild brothers' ace collector, Jordan. Now entomologists are hypothesizing the existence of an even larger moth to pollinate the even larger, 16-inch orchid.

It stands to reason, they say, that a 16-inch tubed orchid would require a giant moth with a 6-inch wing span and a 15-inch proboscis. The nocturnal moth probably sits on the flower's edge and

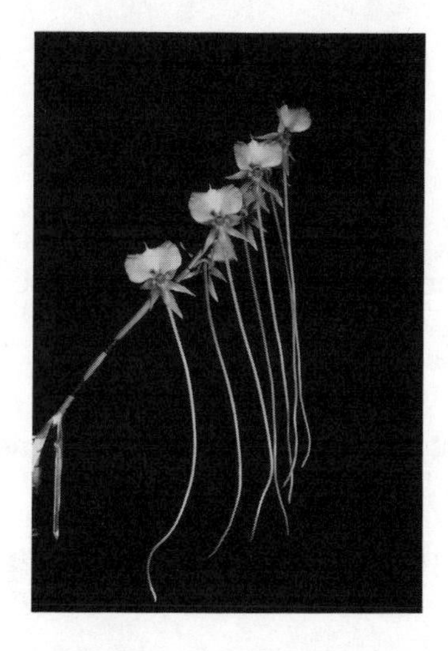

Madagascar orchids, the object of the moth's attention.

In this drawing of the predicted Madagascan moth, its 15-inch proboscis is uncoiled to pollinate a Madagascar orchid's 16-inch-deep nectar tube.

gingerly uncoils its tongue to probe the tube's inside. While there it would collect pollen for later distribution to other flowers. The orchid and the moth probably evolved together; as the flower's tube lengthened, so did the moth's proboscis. Eventually neither could survive without the other. So far, no giant moth has been found.

▶168 **When naturalist James Arnold discovered the largest flower in the plant kingdom in 1818, he wrote, "To tell you the truth, had I been alone, and had there been no witnesses, I should I think have been fearful of mentioning the dimensions of this flower, so much does it exceed every flower I have ever seen or heard of. It measured a full yard across. . . ." What had Arnold discovered?**

Rafflesia arnoldii, a parasitic plant of the jungles of southern Sumatra in Indonesia. The plant reduces parasitism to its bare essentials. It consists only of its outrageously large flower and a weblike system of bridges for transporting water and nutrients from its host.

▶169 **The tropics and subtropics and particularly their rain forests are the most biologically diverse regions of the world. They contain an estimated _____ percent of all flowering plants, ferns, and other vascular plants. (Fill in the blank.)**

A. 68.
B. 32.
C. 24.

A, 68. Of roughly 250,000 vascular plant species known, about 170,000 grow in the tropics and subtropics. Vascular plants represent 99 percent of all land vegetation.

▶170 **What three nations have more plant species than any other countries in the world?**

Three Andean countries: Colombia, Ecuador, and Peru. Although they occupy only 2 percent of Earth's land surface, they accommodate more than 40,000 species.

▶171 Although plants have served medicinally for thousands of years, new drugs are still being discovered in the world's flora. What is the plant that helps cure more than half the victims of Hodgkin's disease and 80 percent of the children with acute lymphocytic leukemia?

The rosy periwinkle of Madagascar. A relative of Vinca rosa, or myrtle, periwinkle produces two alkaloids, vinblastine and vincristine. They are used together with other drugs to fight cancer. Although the rosy periwinkle originated in the tropics, drugmakers farm it today in greenhouses.

▶172 What rare plant is the source of an antitumor treatment for ovarian, breast, and lung cancers?

The Pacific yew (*Taxus brevifolia*). Bark from the tree produces taxol, an antitumor treatment for ovarian, breast, and lung cancer. Pacific yews grow in the moist underbrush of old northwestern forests. Until the drug can be manufactured synthetically, three trees must be killed and stripped of their bark to treat one patient. Chemists are working on ways to synthesize the drug.

Another possible source of taxol is a fungus found in 1991 growing under the bark of a yew tree in an old-growth cedar forest in northern Montana. The fungus, like its host, produces taxol. The taxol fungus *Taxomyces andreanae,* is not the only parasite that produces the same chemical as its host plant. Rice plants, for example, make a hormone that regulates their growth and development; a fungus growing on the rice produces the same hormone.

▶173 Abraham Lincoln's beloved mother, Nancy Hanks, died of poisoning when Abe was seven years old. What killed her?

"Milk sickness," caused by tremetol, an alcohol from the white snakeroot plant (variously known scientifically as *Eupatorium rugosum* or *Ageratina altissima*). Many other pioneers also died from drinking the milk of cows that had eaten snakeroot while browsing in rich woodland. In 1828, two women identified snakeroot as the source of milk sickness: midwife-physician Anna Pierce and an Indian known only as Aunt Shawnee. Their warnings were ignored, however, until 1905 when scientific experiments proved definitively that snakeroot was the culprit.

▶174 What is said to be the most abundant organic chemical on Earth?

Cellulose, the essential component of all plant cell walls. On average, about a third of all vegetable matter is cellulose. Fifty percent of wood and 90 percent of cotton are cellulose.

▶175

Under the spreading chestnut-tree
The village smithy stands;
The smith a mighty man is he
With large and sinewy hands.

—*Henry Wadsworth Longfellow*

Horse-drawn carriages ended a number of promising careers in blacksmithing. But what happened to the chestnut?

In what one expert called "the greatest ecological tragedy in recorded history," several billion chestnut trees along the eastern seaboard died of chestnut blight early in the twentieth century.

The chestnut was one of the most economically important hardwood trees in the East. It provided rot-resistant timber; delicious nuts for food, livestock feed, and flour; and bark and wood for tannin, which is used in leather making.

The blight arrived in the United States with some contaminated chestnut trees that had been imported from the Far East by botanists at the Bronx Zoo in New York City. American varieties had no resistance to the bright orange fungus that girdled their trunks and cut off nutrients and water. In less than 50 years, the blight had spread 25 miles a year up and down the coast and killed several billion chestnut trees.

Using recombinant engineering, scientists have developed DNA that contains a weak strain of the blight, discovered in surviving trees in the Midwest and Europe. Exposing young chestnut sprouts to the weaker strain may inoculate them against the more virulent variety.

▶176 What is the minimum number of plant species needed to feed the world's population?

The traditional answer is between 7 and 30, but new studies suggest that at least 100 or 200 are needed to ensure genetic diversity and vitality and to improve food supplies. Approximately 100 plant species provide 90 percent of the plant foods consumed in 146 countries. Seventy-five of the species supply at least 5 percent of the plant food supply in one or more countries. South Koreans, for example, get 12 percent of their plant food from Chinese cabbage, and taro accounts for 18 percent of the Samoan plant diet.

Earth Science

▸177 For sheer, unadulterated destruction and horror, nothing beats "The Great Dying." What was it?

A close brush with total extinction, when up to 96 percent of all plant and animal species on Earth were destroyed. The mass extinction occurred at the end of the Permian period 250 million years ago. During the relatively short period of a few million years, trilobites and most shallow marine life, giant club mosses, giant horsetails, and many reptiles, among others, were destroyed.

The Great Dying may have been caused by enormous volcanic eruptions that occurred intermittently over less than a million years. Eventually, they covered several million square kilometers of north central Siberia with flood basalt. Clouds of volcanic dust and sulfuric dioxide gas billowed into the upper atmosphere, where the dust and newly formed sulfate aerosols would partly block Earth's sunlight. As temperatures plummeted, polar ice caps may have accumulated enough of Earth's water to cause sea level to fall by more than 200 meters. Eventually the sulfur-bearing compounds in the atmosphere would form sulfuric acid, and acid rain would have fallen on the exposed seafloor.

In comparison, the most famous mass extinction, which exterminated the dinosaurs, is rated a mere fifth in overall destruction. That extinction occurred 65 million years ago, at the end of the Cretaceous period. Currently, the favored theory is that this extinction was caused by one or more asteroids or comets crashing into Earth, although major basaltic eruptions also occurred in India at this time.

▶178 **"Acid flush" is:**

A. Toilet bowl cleaner.
B. Pollutants added to soil and water by melting snow.
C. A dermatological condition.

B, pollution. Falling snow collects air pollutants that are stored over the winter on the ground in snow packs and then suddenly released into soil and streams during a thaw. Scotland experienced black snows almost as acidic as vinegar; the snow was black with carbon particulates from Eastern Europe. Acid flush also may be responsible for the fish kills that often follow spring thaws in Scandinavia.

▶179 **Droughts in Africa benefit:**

A. Miami.
B. Amazonian rain forest.
C. Plankton.
D. The Caribbean.
E. Europe.
F. All of the above.
G. None of the above.

F, all of the above. Winds carry millions of tons of soil from drought-stricken Africa to the Atlantic Ocean, North and South America, and to Europe. The dust fertilizes the ocean's plankton and krill, providing them with up to half of their iron requirements. Rich breeding grounds of fish and other marine life coincide with the main dust routes out of Africa. The eastern Amazon is fertilized with more than 13 million tons of African soil each year, and African dust has been found 1,000 miles from the Brazilian coast. Each spring, Africa's soil mixes with ozone and smog over the South Atlantic as farmers and herdsmen of Brazil and southern Africa use fire to clear vast areas.

▶180 Deep mines are extremely warm, and Earth's outer core is thought to consist of melted iron. Why is the inside of the planet so hot?

Radioactive uranium inside Earth *and* heat left over from Earth's formation. As the uranium decays, it gives off heat that helps melt Earth's outer core of iron. Radiation from the uranium is absorbed by Earth's mass.

The "primordial heat" was generated when Earth formed and when the core and mantle separated; both processes were complete by 4.5 billion years ago. Before the plate-tectonic revolution in geological science, twentieth-century scientists believed that all the heat from Earth's interior was caused by radioactivity.

Heat has been escaping through Earth's outer skin for billions of years. An area the size of four football fields releases 1 kilowatt of heat each day. Altogether, about 42 trillion watts of energy continuously escape Earth's surface.

▶181 The longest distance traveled by any audible, airborne sound is approximately 4,600 kilometers (2,760 miles). What created the sound heard partway round the world?

The volcanic destruction of the Indonesian island of Krakatau, in August 1883. Its most violent explosion, which had an estimated force of 100 megatons to 150 megatons of TNT, propelled ash an estimated 80 kilometers (50 miles) into the air. Sound like distant cannon fire rolled south across Australia, north to Singapore, and westward 4,600 kilometers (3000 miles) to Rodgriguez Island in the Indian Ocean. The explosion created airwaves that bounced seven times back and forth around the world. The waves took 19 hours to reach the opposite end of the world near Bogota, Colombia. Then they bounced back and forth across Earth's surface six more times! Tsunamis, giant water waves, were recorded as far away as South America.

▶182 What are the two hottest years on record?

1990 was the warmest year on record, and 1991 came in second, according to British and American records kept for the past 140 and

111 years, respectively. Moreover, these two warm years came right on the heels of the 1980s, definitely the warmest decade of the twentieth century.

▶183 **What percentage of the world's ice is in Antarctica?**

A. 25 percent.
B. 50 percent.
C. 90 percent.

C, 90 percent. Antarctica is a dry continent, in parts a virtual desert. The average annual snowfall equals much less than a meter of ice. But because Antarctica is so cold, little of its surface ice melts. As a result, parts of the Antarctic Ice Sheet, which covers 99 percent of the continent, are more than 4 kilometers (2.5 miles) thick. In places, the weight of the ice actually depresses the bedrock. Antarctica does not lose ice because of melting; instead icebergs break off, "calving" into the sea.

Most of the rest of the world's ice, the remaining 10 percent, is in Greenland's ice sheet. Unlike Antarctica, some of its ice does melt in the summer.

▶184 **And now there came both mist and snow,**
And it grew wondrous cold;
And ice, mast-high, came floating by
As green as emerald.

—Samuel Taylor Coleridge
"The Rime of the Ancient Mariner"

Icebergs are supposed to be blue and white, made from fresh water that has frozen and calved from a glacier. Did Coleridge have his facts wrong? After all, he wrote his poem in the early nineteenth century.

No, Coleridge was right. One in a thousand icebergs in Antarctica is emerald green. Emerald icebergs form when seawater freezes to the bottom of floating ice shelves. In places, they plaster a layer thicker than a football field is long onto the shelf. When a piece of shelf snaps off and capsizes, the marine side of its ice is exposed.

The ice looks green because yellow plus blue makes green. The yellow comes from the yellowish brown remains of dead plankton dissolved in seawater and trapped in the ice. And the blue is present because, although ice reflects virtually all the wavelengths of visible light, it absorbs slightly more red wavelengths than blue.

Green icebergs exist only in Antarctica. The Northern Hemisphere never gets cold enough. Only in Antarctica can giant ice shelves form, their bases remaining in contact with the ocean's water for centuries.

▶185 **Where is the world's biggest waterfall?**

A. Denmark Strait.
B. Angel Falls, Venezuela.
C. Guaíra Falls, Brazil and Paraguay.

A, the Denmark Strait, a giant undersea waterfall between Greenland and Iceland. It descends 3.5 kilometers (2 miles) and carries 5 million cubic meters of water a second. Most ocean cataracts are caused by convection: the transfer of heat by moving fluid. The Denmark Strait is composed of cold, dense water from the Norwegian Sea that flows from the Iceland-Greenland Rise 650 meters below sea level into the ocean depths. Water flowing out of the Strait continues south for thousands of kilometers along the continental slope of North America. It retains its identity as a distinct chemical entity at least as far south as the tropics.

In comparison, Angel Falls, the tallest waterfall on Earth's surface, is less than 1 kilometer (half mile) high. The Guaíra Falls has the largest average flow of any waterfall on the planet's surface but carries only about 13,000 cubic meters per second. The Denmark Strait's flow is 25 times larger than that of the mighty Amazon River.

▶186 **The Mediterranean Sea is warmer than the Atlantic and loses more water through evaporation than it gains from rainfall and river runoff. When the Mediterranean flows through the Strait of Gibraltar into the Atlantic, does its warmer water rise to the surface?**

No. Because of evaporation, the Mediterranean is saltier, and hence denser, than Atlantic Ocean water. The difference in density between the Mediterranean and the Atlantic moves water in and out through the Strait of Gibraltar. The denser Mediterranean water flows out of the Mediterranean through the Strait of Gibraltar, sinking over the continental shelf into the Atlantic as a giant, underwater waterfall. In turn, fresher and lighter surface water from the Atlantic Ocean flows in through Gibraltar to the Mediterranean. There it evaporates, becomes denser, and begins the cycle anew.

The Strait of Gibraltar Cataract, driven by salinity differences, is different from most giant underwater cataracts. They are caused by temperature differences between warm and cold bodies of water.

Flows from marginal seas like the Mediterranean, Red, Greenland, and Norwegian seas help modify the character of deep oceans.

▶187 **An eddy is a ring of swirling water that separates from an ocean current and twirls off on its own. The North Atlantic has about 1,000 eddies approximately 200 kilometers (120 miles) apart. Many of them are composed of warm water revolving around a cold core. Eddies move at about 75 centimeters (2½ feet) per second and live for months or even years. If that's an eddy, what's a meddy?**

A whirlpool of salty water shed by water from the Mediterranean Sea after it has entered the Atlantic Ocean. Because evaporation is greater than rainfall in the Mediterranean, its water is saltier than the Atlantic Ocean. Meddies of warm and salty water can be 100 kilometers across, live seven years or more, and travel halfway across the Atlantic. Meddies and eddies help mix warm and salty water into the Atlantic at large. They are tracked across the ocean by freely drifting floats.

▶188 **Where can the world's best seasoned firewood be obtained?**

In a mummified forest on Axel Heiberg Island, a mere 680 miles from the North Pole. The 45 million-year-old wood is dry and a bit compressed but is otherwise undamaged. It has its original color and flexibility, and it burns. During the Eocene epoch, one of the warmest periods of the past 100 million years, the North Pole was warm, swampy, and subtropical, much like the Florida Everglades today.

Crocodiles, alligators, turtles, snakes, and lemurs lived among 100-foot-tall dawn redwoods, swamp cypress, cedar, pine, birch, alder, katsura, and other trees. A nearby river flooded periodically and covered the forests with silt so fine that no minerals could reach the wood to petrify it. Only a few hundred feet deep, this soil was not heavy enough to compress the wood into coal. More recently, icy temperatures have kept the surface wood frozen.

▶189 What is the earliest recorded evidence for life on Earth?

3.5-billion-year-old stromatolites near Sarasota Springs, New York.

Stromatolites in western Australia. These fossilized structures, likely formed by microbes, are between 3,450 and 3,550 million years old. Although stromatolites flourished for 2,000 million years long before plants or animals developed, they still survive in a few places, such as Yellowstone Park and western Australia.

Stromatolites look more like concrete cauliflowers than anything else. Approximately 30 centimeters high, they are columns and mounded domes of limestone sediment. They are built by microorganisms (like blue-green algae *cyanobacteria*) that secrete mucus to trap calcium carbonate precipitate from ocean water.

Since they are active only during the day, stromatolites leave a daily record of the sun's position. A stromatolite living about 850 million years ago recorded 435 days in one year. With a 20.1-hour-long day, the mound provides evidence that Earth's rotation rate is slowing, probably because of tidal "friction."

▶190 Although living organisms have inhabited Earth for nearly 4 billion years, large animals did not emerge until 600 million years ago. What took them so long?

Large animals require more oxygen for metabolic activity than existed earlier in Earth's development.

Virtually all free oxygen comes from photosynthesis, in which plants use the sun's energy to convert carbon dioxide and water into sugars and oxygen. When plants die and decompose, their carbon recombines with atmospheric oxygen. To accumulate excess oxygen, decomposing plants must be prevented from recapturing oxygen. About 600 million years ago, major volcanic and climatic changes buried so much plant material deep under ground and sea that oxygen could accumulate in Earth's atmosphere.

▶191 Superhot plumes of water bubbling up through holes in the ocean floor are home to thriving communities of anemones, shrimp, crabs, clams, mussels, limpets, and tube worms. These seafloor oases exist so far from the sun that their organisms cannot possibly depend on photosynthesis to convert energy produced by the sun. So what do they live on?

Chemical-guzzling bacteria. Water heated under pressure to more than 600°F jets up through fissures in the seafloor. The water is rich in sulfur compounds leached off hot volcanic rocks underground. Bacteria in nearby warm water oxidize these sulfur compounds to produce the energy they need, and specially adapted creatures filter out the bacteria for food.

The hot springs form plumes of water up to 150 meters high. Their sulfur compounds form when water, heated by underground volcanic activity, reacts with rock. The springs were discovered in the Galapagos Rift of the Pacific Ocean, an area where the earth's tectonic plates are spreading apart. Others have been discovered in the Atlantic 1,800 miles due east of Miami in the Atlantic Ocean.

Other seafloor oases grow around cold plumes of especially salty water and in regions where hydrocarbons seep from the seafloor. A total of 223 invertebrate species have been identified, most of them unique to their particular vent. Hot water vents may live only 50 years. What happens to their resident populations when the vents die? And how do new vents get populated? No one knows.

▶192 What island region has the greatest concentration of active volcanoes on Earth?

A. The Philippines.

B. Easter Island.

C. Indonesia.

B, Easter Island, where 1,133 seamounts (submerged mountains) and volcanic cones were discovered in 1992. Many of the volcanoes rise more than a mile above the ocean floor. Some are almost 7,000 feet tall, although their peaks are still 2,500 to 5,000 feet below the sea's surface. In some places, the volcanoes form 300-mile-long mountain ranges underwater. Sonar scanning devices discovered the region, about the size of New York State, near the East Pacific Rise. The East Pacific Rise parallels South America to the west of Easter Island.

Scientists estimate that two or three of the volcanoes could be erupting at any one time. Their activity may precipitate El Niño, which occurs when the giant high-pressure system centered near Easter Island drops and triggers changes in Pacific water currents, weather, and fishing patterns.

More than 60 percent of all known volcanoes occur along the rim of the Pacific Ocean, and 90 percent of all volcanic activity is on the ocean floor.

▶193 Earth's crust is composed of about two dozen giant slabs or plates ranging in thickness from 50 to 80 miles. Some of these rigid plates are moving apart. How fast do the fastest-separating plates move?

Eight inches a year; the Pacific and Nazca plates at the East Pacific Rise, an undersea ridge running through the Pacific Ocean west of South America and Easter Island. Other plates move about 5 inches per year.

▶194 **Most volcanoes explode at the edges of continental plates that are moving apart or pressing on top of one another. But some volcanoes are "misplaced." They exist where none should be: in the middle of plates. Hawaii, Yellowstone National Park, and Reunion Island in the Indian Ocean are isolated volcanic regions located far from the borders of any continental plate. (In the case of Yellowstone Park, hot springs and geysers replace full-blown volcanoes.) What causes misplaced volcanoes?**

Hotspots on Earth's surface. They are warmed by plumes of hot rock rising a few meters per year through vertical chutes from deep within Earth. The hotspots are relatively fixed in position as the continental plates glide overhead. A plume under Hawaii, for example, created the volcanic island of Oahu and, millions of years later, the island of Hawaii; thanks to shifting continental plates, the two islands are now 120 miles apart. Approximately 40 hotspots are currently scattered across Earth's surface.

The plumes shooting up through Earth's crust also create bulges in the ocean floor. Both Iceland and Hawaii are on top of such bulges. The water around them is 2,000 feet shallower than anything for hundreds of miles around.

▶195 **Where is the deepest soil on Earth?**

In north central China. Loess deposits in the heartland of ancient China can be more than 600 feet deep. Heavy winds blow the fertile silt in from central Asian deserts. China's early culture blossomed on the loess plateau, thanks to its extraordinary fertility.

▶196 **What do cattle, marshes and bogs, rice paddies, forest fires, burning grassland, garbage dumps, termite mounds, and asphalt roads have in common?**

They make the greenhouse gas methane that contributes to global warming. In fact, they make about 500 million tons of it every year. Methane is produced when bacteria break down organic matter in an oxygen-free atmosphere, for example, in intestines or underwater. Termites alone may produce 5 million tons of methane yearly while sunlight on asphalt in the United States is responsible for another 5 million tons. The roots of rice plants transport methane from mud underwater up into the air.

▶197 **Besides milk, cream, butter, cheese, and calves, a cow produces about 7 ounces of ______________ each day. (Fill in the blank.)**

Methane gas. Multiply that by approximately 1,300 million head (or rear) of cattle, and you get 100 million tons of methane yearly, enough to warm the planet Earth.

The world's population of cattle has doubled in the past 40 years until today there is roughly one head of cattle for every four people. Bacteria that break down cellulose in the animals' intestines convert between 3 percent and 10 percent of their food to methane.

By 2050, methane may surpass carbon dioxide in importance as a greenhouse gas. Each molecule of methane traps 25 times more solar heat in the atmosphere than a molecule of carbon dioxide does. The concentration of methane in the air is increasing four times faster than the concentration of carbon dioxide. And 50 million more tons of methane enter the atmosphere annually than leave it.

▶198 **What country produces the most methane?**

India, with its millions of cattle and miles of rice paddies. But if the world warms, bogs and tundra in Canada, Siberia, and Scandinavia will produce more. Vast amounts of ancient methane stored in the ground will be released as the permafrost layer warms and thaws.

▶199 **A theory about hotspots, traps and steps, molten rock, and dinosaurs links two types of rare but giant catastrophes. What are they?**

Basalt floods and the mass extinction of species. About a dozen times in the past 200 million years, continental plates have rifted apart over a plume of magma rising from Earth's mantle. Enormous floods of lava welled up, covering large portions of the globe with basalt (a greenish-black igneous rock) a mile or more deep.

Basalt floods cover portions of South America, Namibia, Washington, Oregon, and Idaho, northern Great Britain, Ethiopia, the

North Atlantic floor, Antarctica, Siberia, and India. The most famous are the Deccan Traps, which formed 66 million years ago as the dinosaurs were dying out. The Traps, named for the Dutch word for "step," cover a third of the Indian peninsula and are up to 1.5 miles thick. Dust and gases from the volcanic explosions could have produced enough acid rain and blocked enough sunshine to kill the dinosaurs' food supply.

▶200 **Imagine the ultimate sci-fi horror: Erupting volcanoes under the Antarctic ice cap melt enough ice to make the ice sheet collapse, submerging New York City, Los Angeles, New Orleans, Tokyo, Hong Kong, Bangkok, and other major population centers. Is this scenario possible? Or is it pure fantasy?**

Potentially possible. At least one active volcano lies under more than a mile of ice in the globe's coldest region in western Antarctica. Conically shaped like Mt. Fuji, the volcano's peak rises 700 yards above its surrounding bedrock. The ice above it is slightly depressed, forming a circular hole 50 yards deep and 3.5 miles across. If the heat associated with erupting volcanoes melted enough ice from the base of West Antarctica's ice sheet, it could collapse, causing sea levels worldwide to rise 6 meters.

The Antarctic ice cap has expanded and shrunk over geologic time. Large portions have disappeared into the sea and then re-formed over millions of years. The current West Antarctic ice sheet may have started growing between 600,000 and 2 million years ago.

▶201 **A mysterious object exploded over Siberia with the force of 1,000 Hiroshima-sized atomic bombs on June 30, 1908. It exploded 5 miles up in the air, leveled hundreds of square miles of trees, and started fires for miles around. A pressure wave from the explosion was recorded around the planet. No piece of the object has ever been found. What was it?**

A. A comet.
B. An asteroid.
C. A tiny black hole.
D. An alien spaceship.

B, an asteroid. This is the accepted explanation, although the others have been discussed, too. According to a 1993 mathematical simulation of the Tunguska explosion, the object was a stony asteroid with a radius of 100 feet. A smaller asteroid would have exploded higher in the atmosphere, and a larger one would have struck Earth. Tunguska-type collisions occur every 300 years or so. More than a million asteroids 100 feet or more in diameter have orbits crossing Earth's path.

▶202 **What is the chemical composition of seawater?**

Seawater is a chemical soup containing nearly all the known elements. Thanks to the discovery of deep-sea vents, scientists now realize that the chemical composition of ocean water is constantly changing, according to volcanic and venting activity on the ocean floor. Thus, like all great soup recipes, the formula for seawater changes according to what is locally available.

▶203 **Water bubbles up from below Earth's crust in undersea hot springs, geysers, and smokers. When the metallic solutions in the geysers precipitate, they can form chimneylike structures up to 75 feet tall and black smoke rich in metal-sulfide particles. The hottest of all seafloor hot springs is the black smoker. How much water is involved in such hot springs and geysers on the seafloor?**

A. A trickle.
B. A river-sized amount.
C. An ocean's worth.

B, a river-sized amount. Hot springs and smokers are thought to exist throughout the world, near volcanically active islands and near the island chains of the western Pacific. Water constantly seeps through cracks in Earth's surface down into the magma, where it is heated by hot rocks, expands, rises, and leaches minerals from the surrounding rocks. The annual flow of seawater circulating down through these hot rocks and back up into the ocean again is comparable to the amount of water in the world's largest river, the Amazon. It is thought that all ocean waters circulate over these undersea hot rocks every 10 million years. At that rate, all the water in the oceans will have flowed over the hot rocks hundreds of times in the history of Earth.

▶204 Where do the minerals in seawater originate?

A. In rocks on land. Rivers wash them into the sea.
B. Molten rock inside Earth.

A and B. Until deep-sea vents and smokers were discovered, it was thought that all the minerals originated in rocks on land and that they were washed into the sea by rivers. Now it is realized that much of the copper, iron, manganese, rubidium, lithium, potassium, calcium, silica, and zinc in seawater comes from volcanic rocks, dissolved by hot water inside Earth's crust.

▶205 Raised beds are the height of fashion for backyard gardens today. Where were they also popular more than 1,000 years ago?

Raised fields in the Llanos de Mojos in the Bolivian Amazon.

In South America and Mexico, where vast systems of raised beds and water ditches drained fields and provided irrigation, algae fertilizer, and fish. Savannas, now so waterlogged that they are used primarily for cattle pasture, once supported large populations. Bolivian lowlands, now flooded yearly by tributaries of the Amazon, were farmed 1,000 years ago, thanks to hundreds of square miles of raised fields. Other ditch-and-ridge systems have been found in parts of Colombia, Ecuador, Surinam, Venezuela, Mexico, Belize, and Peru. They protected crops from killing frosts, severe drought, and heavy flooding.

▶206 What are the oldest known rocks on Earth?

Granite rocks dating from 3.96 billion years ago. Found in the tundra of northwestern Canada near Great Slave Lake, the rocks suggest that Earth was forming continental crust about 4 billion years ago. The planet is about 4.5 billion years old. Despite their age, however, the granite had evolved from even earlier rock.

As the molten granite cooled, zircon crystals formed and locked inside them tiny amounts of naturally radioactive uranium. Scientists dated the rock by calculating how much of the uranium inside the zircon crystals had decayed radioactively into lead.

Until the discovery of the Great Slave Lake rocks during the 1980s, the oldest rocks known were located in western Greenland. They are about 100 million years younger, however.

▶207 Has anything older than the Great Slave Rocks been discovered on Earth?

Yes, zircon crystals almost 4.3 billion years old are the oldest objects on Earth. Found in Australia, they eroded from their original rock and were incorporated into younger material. Thus, they do not tell us much about conditions on early Earth. The planet began to form about 4.5 billion years ago.

▶208 Where did a world-shaking collision between Earth and an extraterrestrial object create Earth's largest known impact crater?

A. Manson, Iowa.
B. Puerto Chicxulub, Mexico.
C. Popigai, Siberia.

B, Puerto Chicxulub, Mexico. A bull's-eye crater 180 kilometers (112 miles) across lies beneath the village of Puerto Chicxulub on the northern coast of the Yucatán peninsula. The Chicxulub impact may have caused the catastrophe that killed off the dinosaurs and between a third and a half of all marine genuses at the end of the final Cretaceous period 65 million years ago.

▶209 **The location of magnetic north—the north that compasses point to—wanders around, reflecting the turmoil inside Earth's outer core. Variations in Earth's magnetic field suggest that material in the outer core moves several kilometers per year. In contrast, the mantle is a 3,200-kilometers-thick (2,000-mile-thick) shell of solid rock that moves much more slowly, at the speed of continental drift. Hot and under enormous pressure from the layers of rock above it, the mantle rotates in slow motion like a viscous liquid a few centimeters per year.**

This view of Earth's interior is new. During the nineteenth century and most of the twentieth century, geologists believed it to be static and unchanging. Who first suggested that changes in magnetic north are caused by the slow rotation of a magnetized body inside Earth?

Edmond Halley in 1692.

▶210 **Some giant landslides race for miles. These mammoth avalanches move so fast and so far that they seem to violate the laws of friction. Many slide horizontally ten times farther than they fall. The Blackhawk slide in the Mojave Desert east of Los Angeles spread 5 miles at a speed of 70 miles per hour. A prehistoric landslide on the seafloor near Hawaii traveled 120 miles, thirty times longer than its fall. One of the largest landslides on planet Earth, it set off gargantuan tsunamis. How can giant landslides travel so far so fast?**

A. They "fly" on a layer of compressed air.
B. They ride on a layer of rock melted by friction.
C. They flow like water.
D. Noise vibrations lubricate their slide.

All four theories have been proposed, but scientists still argue which is right. Ultimately, computer simulations may provide the answer.

▶211 **Mysterious circles of stones, Zenlike fields of striped rock gardens, and uniform polygons outlined in ice—all are soil patterns formed in frozen ground. Whether in the Arctic or on the slopes of Hawaiian volcanoes, the patterns may be repeated over and over again for kilometers around. Who or what made them?**

A. UFO's.
B. Irish monks, tenth century.
C. Frost.

Stone circles in the Arctic.

C, frost. Soil patterns, as such formations are called, occur where all but the top meter of the ground is permanently frozen. Patterns vary in size from Arctic circles that can be 3 to 6 meters across and up to 50 centimeters high to ice wedge polygons 10 meters across. The circles are formed by soil convection over tens of thousands of years. As water trapped in fine soils freezes, the soil expands and wells up in convection patterns.

On hillsides, ice needles form on cold, clear nights and push grains of dirt up to the surface. When the needles thaw, the finer grains fall back down into the earth, and the coarse pebbles roll downhill. This action, repeated over long periods of time, eventually forms alternating stripes of fine and coarse soil.

▶212 No two snowflakes are ever alike. Right?

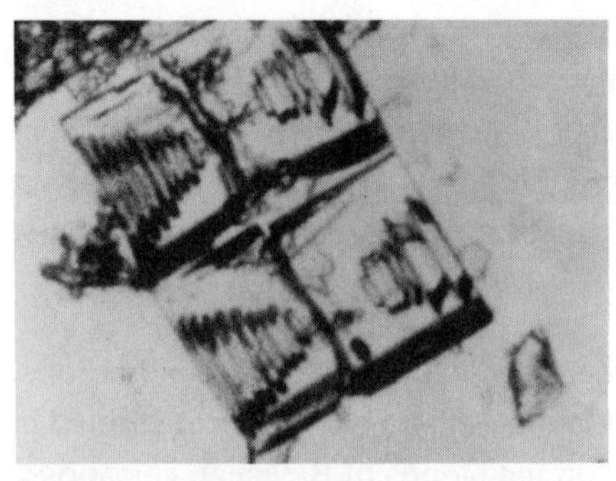

The first two identical snowflakes. Each of the two side-by-side crystals is shaped like a thick, hollow column with the skeleton of a vaselike structure inside. The pair probably grew attached to each other.

Wrong. The first two identical snowflakes were collected on an oil-coated glass slide extended from an airplane flying over Wisconsin on November 1, 1986. The plane was flying between two layers of clouds at 6,134 meters (20,000 feet) where the temperature was −22°C (−9°F). The pair of crystals had grown together side by side. And if they weren't scientifically identical, the experts said they certainly looked "alike."

The Molecules of Life

▶213 **Children with Prader-Willi syndrome, first described in 1956, are mildly to moderately retarded and have little muscle tone. Although they are insatiably hungry and become obese, they have tiny hands, feet, and gonads.**

Children with Angelman syndrome, described nine years later, are severely retarded and move their bodies jerkily and repetitively. They often have seizures and incoherent speech.

An almost unimaginable condition connecting the two syndromes was discovered in the late 1980s. What is it?

Both groups of children lack the same piece of genetic material on chromosome 15. But one condition is transmitted by the child's father, and the other by the mother. Prader-Willi patients lack their father's genetic material while Angelman patients lack their mother's. For the first time, human geneticists realized that one genetic defect could cause two different diseases, depending on whether the father or the mother transmitted the defect.

▶214 **What's the difference between the smell of spearmint and a whiff of caraway or dill seeds?**

Chirality. Chefs may talk about bouquet, fragrance, and a soupçon of "je ne sais quoi," but the primary difference is molecular. Their molecules are mirror images of each other. They are composed of the same atoms ($C_{10}H_{14}O$) and have the same structure. But the molecule for spearmint's odor is left-handed, so to speak, while the molecule for the scent of caraway and dill is right-handed. Thousands of chemical compounds exist as mirror-image twins.

Such pairs, called chiral (KYE-ral) or handed-molecules, can have very different properties. One partner may cure a disease while another is poisonous. One twin can be produced without the other. For example, the drug l-DOPA is prescribed to help alleviate the symptoms of Parkinson's disease, while its right-handed twin is mildly poisonous.

Strangely, the amino acids of virtually all living things on Earth are left-handed (called L). Extraterrestrial objects like meteorites, on the other hand, are mixtures of right- and left-handed molecular twins. At the same time, the body digests right-handed glucose sugar (called D) but not its sweet left-handed twin. Perhaps someday dieters, whether left- or right-handed, will munch southpaw glucose without any caloric effect at all.

▶215 What's more like an elephant—a mushroom or a pine tree?

The mushroom, according to a new theory of their evolution. Both animals and fungi may have descended from a protist, a single-celled part fungus/part animal that spent some of its life like a sperm and some with a stiff cell wall like a fungus. A computerized mathematical model created a genetic family tree for fungi stretching back 1 billion years. If true, the theory could explain why it is so difficult to treat fungal diseases without also harming their animal hosts. Physiologically, the two are simply too close. However, the theory may be news to mycologists, the people who study fungi. They tend to work in botany departments rather than zoology. Musical chairs may be in order.

▶216 How big do bacteria get?

Roughly half the size of a poppy seed. *Epulopiscium fishelsoni,* the world's largest bacterium, is a bizarre organism. It's so big and yet so simple. It has a million times more volume than *Escherichia coli*, the world's most studied bacterium. In fact, *Epulopiscium* is so large that it's visible to the naked eye. It took biologists several years before they could even tell what kind of cell *Epulopiscium* is. Traditionally, cells are divided between the large and complex eukaryotes that have cellular nuclei of genetic material, and the small and simple prokaryotes without nuclei. The problem was that, although *Epulopiscium* is large and has a rather complex internal structure, it lacks a nucleus. DNA analysis in 1993 finally proved *Epulopiscium* to be the world's largest prokaryote *and* the world's largest bacterium.

Epulopiscium is a member of the *Clostridium* family, some of whose members cause the fatal diseases, botulism and tetanus. But don't worry about getting sick from *Epulopiscium.* So far, it's been found only in the gut of a Red Sea surgeonfish.

▶217 In genetic engineering or transgenics, foreign DNA is injected into an embryo to become part of the genetic makeup of the developing organism and its succeeding generations. In what organism was the technique first developed?

A. Fruit flies.
B. Nematodes (roundworms).
C. Amphibians.
D. Mice.

Surprisingly, in D, mice. Transgenics is a rare case of Mammals First. Most studies in genetics and development biology have been done first in simple species. Investigators trying to correct a genetic defect, mouse dwarfism, injected growth hormone genetic material into mice in 1982. The results, announced in headlines around the world, was Supermouse. Supermice grew faster and bigger than normal mice. Later, the investigators were able to actually correct dwarfism in mice. It was the first genetic defect corrected in a transgenic animal.

▶218 Animal domesticators, like gentlemen in Broadway musicals, preferred blondes. Why?

Domesticated species of animals are smaller, lighter in color, less brainy, and more relaxed than their wild relatives. The melanins that darken animal hair are biochemical cousins of the neurotransmitters that make the animals sensitive to their surroundings. Thus, the darker the species, the more jittery it is likely to be. And the lighter the species, the more docile it may be.

As for their intelligence, domesticated animals come from species that are more primitive and less intelligent than their wild relatives. And the brain size of domesticated animals shrank even further as a byproduct of their domestication.

▶219 Who are the chimpanzee's closest relatives? Gorillas? Or human beings?

Human beings, according to a molecular comparison of DNA from the three species. Researchers analyzed a 7,000-nucleotide sequence from the same stretch of DNA in gorillas, chimpanzees, and human beings. From that, they inferred what the ancestral gene common to all three looked like. Then they figured the simplest way that the three could have evolved and branched off their common family tree.

According to that, gorillas diverged from the common ancestor first, followed by the chimpanzees and then human beings about 5 to 10 million years ago.

The problem? Nothing in current anatomical studies suggests that humans were ever knuckle-walkers like gorillas and chimpanzees. Stay tuned for the next episode.

▶220 What species of wild animal has as little genetic diversity as an inbred strain of livestock or laboratory mice?

The cheetah (*Acinonyx jubatus*), the world's fastest land animal. The cheetah's lack of genetic diversity makes it both susceptible to disease and infertile in captivity. Only 15 percent of cheetahs who were born wild have ever reproduced in captivity, and their young have a high mortality rate.

Fewer than 20,000 cheetahs remain in sub-Saharan Africa. They used to roam North America, Europe, Asia, and Africa. DNA analysis suggests that they suffered a population crash approximately 10,000 years ago, about the same time that 75 percent of all large mammals in North America, Europe, and Australia became extinct. Human hunting or an environmental catastrophe have been considered possible causes. Cheetahs are an endangered species, and rearing them in captivity has proved difficult.

▶221 How much of the human genome is composed of genes?

A. 98 percent.
B. 50 percent.
C. 2 percent.

C, 2 percent. The remaining 98 percent, traditionally called "junk DNA," remains a mystery. Each meter of DNA contains three billion pairs of bases or nucleotides. Only two percent of them encode genetic data. The remains may help control the genes in some way: assist the meter-long molecules of DNA to pack themselves into the tiny nuclei of the cells; serve as a dumping ground for unused genes; or act as a source for building new genes.

▶222 What do Huntington's disease, fragile X syndrome, myotonic dystrophy, and spino-bulbar muscular atrophy have in common?

All are hereditary diseases caused by "stuttering" genes. In each case, a segment of DNA is repeated too often. Huntington's disease involves the repetition of a gene on chromosome 4. Normally, people have between 11 and 34 repeats of the three-base triplet, cytosine-adenine-guanine. People with Huntington's disease can have between 42 and 100 repeats of the stuttering triplets. And in many cases, the more triplets the patients have, the younger they were when the disease began. Huntington's disease is a neurodegenerative illness that produces uncontrolled movements of the limbs, mental deterioration, and eventual death.

Fragile X syndrome is second only to Down syndrome as a cause of mental retardation among males. Myotonic dystrophy is the most prevalent muscular dystrophy in adults. And spino-bulbar muscular atrophy is one of several spinal muscular atrophies that, together, are the second most common serious neuromuscular disorder among children.

▶223 **Charcot-Marie-Tooth syndrome is an inherited disease in which the nerves of the feet and hands atrophy. Its cause is an entirely new and unexpected kind of genetic defect that caught researchers by surprise. What is it?**

A triple helping of DNA. Instead of deleting or shuffling a sequence of DNA bases, the sequence is repeated. Thus, people with Charcot-Marie-Tooth syndrome get three copies of a tiny portion of chromosome 17 instead of two. A normal chromosome has approximately 130 million base pairs of DNA, but Charcot-Marie-Tooth patients have an extra copy of about 500,000 of them. This triple defect is expected to be involved in a number of other inherited diseases too.

The syndrome affects one in 2,500 people. Although it is a relatively mild disease of the peripheral nervous system, it is progressively debilitating.

▶224 **How small are the smallest viruses?**

Only 18 nanometers (18 billionths of a meter) across. Satellite viruses have such tiny protein shells that they can't hold all the genetic material that the viruses need to replicate on their own. So they parasitize the viruses that parasitize cells. The host virus supplies the genetic code for everything that the satellite virus cannot provide itself.

Satellite viruses are roughly spherical. They are shaped like Buckminster Fuller's geodesic domes or like carbon buckyballs. One of the smallest satellite viruses, satellite tobacco necrosis, has 60 triangularly shaped sides.

In comparison, a herpes virus measures 100 nanometers or 100 billionths of a meter across, and bacteria typically measure 1,000 nanometers. A typical plant cell is 500 to 5,000 times larger than a satellite virus.

▶225 **Traditional vaccines are purified proteins or infectious agents that have been weakened or killed. Injected into a person or animal, they produce antibodies. But isolating enough pure protein can be time consuming, difficult, and even dangerous. Making vaccines against viruses like influenza and human immunodeficiency virus (HIV) is especially challenging. The same virus may be enveloped in a variety of different proteins, each of**

which gives rise to a different antibody that is effective only against its particular protein. What more direct method of vaccination shows promise?

A genetic vaccine. Genetic vaccination would inject genetically altered, naked DNA directly into the cells of living animals. Intramuscular injections or microscopic, DNA-coated projectiles would carry a gene encoded to produce a particular virus protein. Once inside the animal's cells, the gene would replicate the protein, which would then elicit an immune response and antibodies. Thus, instead of producing the protein in a laboratory and injecting it into a person, the gene would produce the protein inside the body.

It is hoped that the method, which has been tested and proven effective in laboratory animals, can be used to vaccinate people against a wide variety of viruses.

▶226

The genetic makeup of modern thoroughbred horses was built on the "Pillars of the Stud Book." What are the pillars, and how strong are they?

The four pillars of modern thoroughbreds were four stallions imported to England in the 1600s. They account for one-third of the genes in the entire population of thoroughbred horses today. Half of their genes come from ten horses. The pillars may be deteriorating though. The thoroughbred's low fertility rate may be the result of 20 generations of inbreeding, coupled with racing regulations that encourage breeding during the three winter months when mares are least fertile.

▶227

What famous molecule is 1 meter long?

The amount of chromosomal DNA in a human egg or sperm cell would be approximately one meter long if it were linked up end-to-end and stretched out. However, most human cells contain 46 chromosomes for a total of about two meters' worth of DNA. (Unlike sperm and egg cells, most cells contain two copies of each chromosome.) The average chromosome contains a DNA molecule approximately five centimeters long.

To squeeze inside the cell, the DNA molecules fold and refold themselves in various ways. Each DNA molecule includes a sequence of bases that contains the genetic information for the organism.

▶228 **"The Fruit Fly of the Plant World" has fewer genes than any other known flowering plant. Thus, its genome is only slightly larger than two other well-known research subjects for geneticists: baker's yeast and *E. coli*. Furthermore, it has few of the repeated DNA sequences that confuse genetics studies of other plants. Add to these charms a six-week life span, ease of cultivation, and thousands of seeds and you have a hot new tool for plant genetics. What is it?**

Arabidopsis thaliana, a foot-tall member of the mustard family. A weed found round the world, it thrives in a laboratory test tube. With only five haploid chromosomes and about 20,000 genes, it is used to bridge the gap between molecular genetics and plant breeding.

Most plant genomes are ten times larger than those of mammals and 100 to 10,000 times bigger than those of a bacteria. But unlike an animal cell, a single plant cell can produce a complete, fully functional organism.

▶229 **What was the first animal to be patented in the United States?**

The Harvard mouse, a white mouse with a human cancer gene, patented in April 1988. The first organisms protected under a standard U.S. patent were oil-eating bacteria developed to clean up oil spills. Patented plants include herbicide-resistant cotton, insect-resistant tobacco, and virus-resistant potatoes.

▶230 **Proteins are crucial to virtually all biological processes. They catalyze reactions, transport and store other molecules, control growth and differentiation, transmit signals, and provide immune protection. Their ability to perform their various jobs depends on the shape they assume when they are folded tightly up. Nature seems to have an almost limitless variety of proteins and protein functions. So how many different types of protein structures are known so far?**

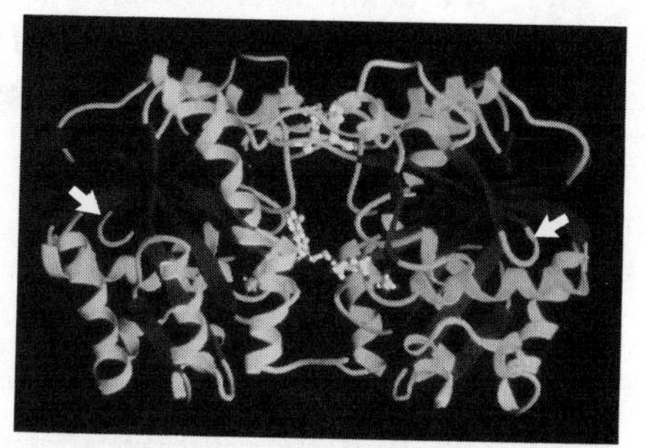

Protein folded into loops (indicated by arrows) and into alpha helices, which look like twisted telephone cords. The protein is part of nitrogenase, which turns nitrogen in air into ammonia in legume plants and nitrogen-fixing bacteria.

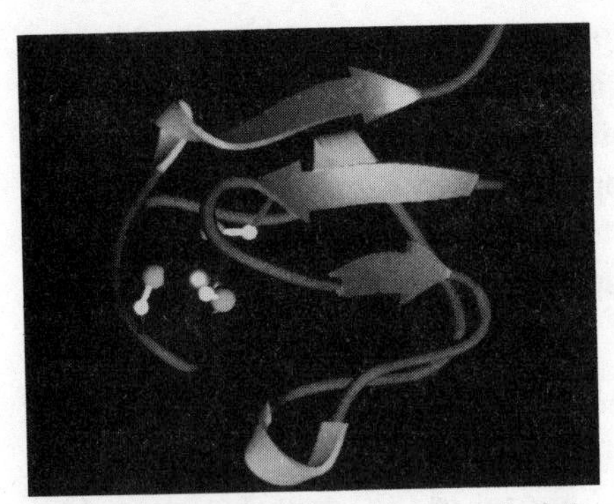

Beta-strands represented as arrows make beta-sheets in a small protein rubredoxin, which may help transport electrons.

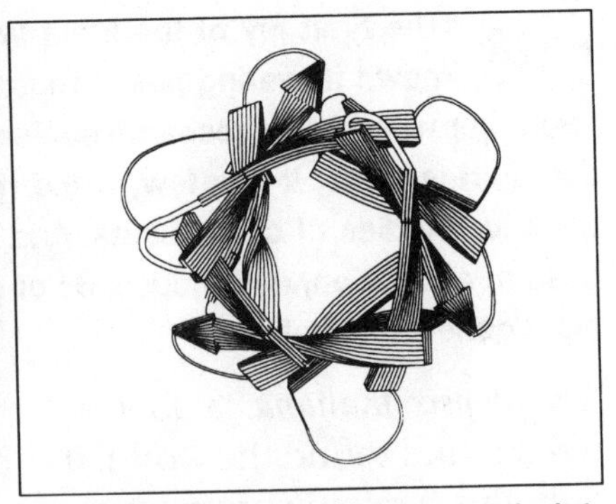

Beta-strands, represented as arrows, make beta barrels and tubes. The beta-barrel is decorated with loops, represented by strings. This protein is the fibroblast growth factor that helps make connective tissue.

A. Fewer than two dozen.
B. Fewer than 200.
C. Fewer than 2,000.

A, fewer than two dozen. Although approximately 1,000 folded proteins have been analyzed, they are all composed of three basic structural elements: alpha helices, beta strands, and loops. These three shapes are combined to make structural motifs like beta barrels and beta sheets. Subtle variations in these basic shapes drastically affect what the proteins do.

▶231 The amino acid sequence of a protein determines its three-dimensional structure. How many of the amino acids have to be in the same sequence for two proteins to have essentially the same three-dimensional form?

A. 30 percent.
B. 50 to 60 percent
C. 80 to 90 percent.

A, 30 percent. Many different sequences seem to code for almost identical structures. Structures are considered similar if they have the same general folds in the core of the protein. Many families of proteins have quite similar three-dimensional structures overall when only 20 or 30 percent of their amino acid sequences are the same. For example, the three-dimensional structures of myoglobin and hemoglobin in humans resemble each other closely. But only 24 of their 141 amino acids are in the same sequence. Other protein families are more similar, of course. Forty percent of the amino acids in the digestive enzymes, trypsin, chymotrypsin, and elastase are in the same sequence.

▶232 **What is the most abundant protein in mammals? What is its job?**

Collagen, which provides structural support for the body. It is the major fibrous element in skin, bone, teeth, tendon, cartilage, and blood vessels. Collagen makes up one-quarter of the total weight of mammals. The protein consists of long fibers made of three protein strands wound around each other. It is the shape of the protein that allows it to perform its job.

▶233 **Besides being proteins, what do insulin, interleukin-8, and antibodies have in common?**

They transmit signals by binding to other proteins. Thanks to their shape, each fits into another chemical like a key into a lock. Many biological processes depend on one cellular component's recognizing another by how well their two structures fit together. The specific shapes of insulin, interleukin-8, and antibodies permit them to bind to their target proteins.

For example, when insulin is released into the bloodstream, it binds to insulin-receptors on the surface of cells to trigger the reaction that makes the cell absorb blood sugar. The insulin binds to the receptors because their two structures fit precisely.

Interleukin-8 is released at the site of inflammations. It spreads and binds to certain white blood cells, which find their way to the inflammation site by moving in the direction of the interleukin-8.

The antibodies of the immune system recognize foreign particles or antigens by binding to them; once the antibody and antigen are locked together, the immune system can attack. However, the antigen and antibody bind together only if their shapes fit together.

▶234 **The transparency of its embryo opens a window on developing life. Individual cells deep inside its embryo are visible as they divide and**

grow. Twenty-four hours after fertilization, the beating heart is visible. What familiar vertebrate became a model for nervous system studies in the 1980s?

The zebra fish *Brachydanio rerio,* favorite of tropical fish hobbyists because they're so easy to grow. The embryos of these tiny, freshwater fish hatch after 48 hours. Fourteen days later, an egg can be a breeding adult.

▶235 **Once upon a time, about 220 to 230 million years ago, a microcommunity of soft-bodied, one-celled animals and plants lived in Bavaria, Germany. Amazingly, many of them looked just like their modern-day descendants. Suddenly, the tiny community was inundated by a wet, gooey substance. The material sealed off the organisms from water, air, and bacteria and preserved them from decay. They became the earliest known fossils of terrestrial, soft-bodied, one-celled organisms. What was the preservative?**

A. Tar.
B. Lake mud.
C. Tree resin.

C, tree resin, fossilized as the gem amber. Among the fossils are cyanobacteria, sheathed algae and fungi, germinating pollen and spores, and fungal spores. They probably lived on the very tree that produced the resin that encased and preserved them.

Most organisms trapped in amber are arthropods. DNA has been extracted from an extinct, 120- to 135-million-year-old weevil that ate Lebanese conifer trees and was, in turn, entombed by the resin from one. It was the oldest fossil DNA ever extracted and analyzed. The oldest plant DNA extracted from amber, also analyzed in 1993, lived 25 to 40 million years ago in what is now Hispanola. It came from an extinct tree, *Hymenaen protera* Poinar. The plant's DNA supports the theory that the plants of Africa and of North and Central America are more closely related than the plants of Africa and South America.

It was tar, however, that preserved the mammoths, mastodons, giant ground sloths, and dire wolves in the La Brea Tar Pits of Los Angeles. DNA has been extracted from the 14,000-year-old bones of an extinct saber-toothed cat (*Smilodon fatalis*) found there. The analysis showed that the saber-toothed cat was closely related to modern cats like lions and tigers. Elsewhere, lake sediments preserved fish scales 200 million years ago.

▶236 **After eating a banana, an exceptionally beautiful quarter horse suffers uncontrollable muscular spasms and temporary paralysis. With a mildly irregular heartbeat and obstructed airways, it topples over and dies. What killed the horse?**

A defective gene. The same gene that produces beautiful musculature in quarter horses also gives them hyperkalemic periodic paralysis. The gene is believed to affect 30,000 to 60,000 quarter horses in the United States. They may all have inherited it from one particular stallion used as a stud during the 1970s and 1980s. Because a test can determine which of the nation's 2.9 million quarter horses carry the gene, the condition could be eliminated in a single generation. Horse owners may not want to do away with the defect, though. Quarter horses are raised as much for show as for racing. And the disease, which affects the horse's ability to metabolize potassium-rich foods like alfalfa and bananas, can be controlled by diet and diuretics.

▶237 **As a result of the Potato Famine in Ireland, 2 million people died and another 2 million emigrated. Irish peasants had subsisted on a diet of potatoes and milk or buttermilk. Provided they could get enough potatoes, the Irish had a fairly balanced diet, thanks to the protein, vitamins, minerals, and calories in the potato. In 1845, 1846, and 1848, however, the crops failed totally. With no money to buy other food, the poor starved. What caused the Potato Famine?**

A fungus and the potato's lack of genetic diversity. Irish potatoes were the descendants of a few potatoes brought to Europe from the Andes of South America. Removed from their native pests and diseases, potatoes flourished in Ireland until the late potato blight fungus struck. Because of their narrow genetic base, the Irish potatoes had no resistance to the fungus and entire crops were destroyed.

In 1970, half the U.S. corn crop was lost to another fungal disease for a similar reason. Most American corn varieties at the time shared one gene that made them susceptible to the disease. New genes from other more resistant varieties were added to strengthen the American types.

Chemistry

▶238 What is the roundest molecule of all?

The "buckyball," or C_{60}. Scientists assumed they knew everything about pure carbon and its manifestations as graphite and diamond. Then, the buckyball was discovered in 1985 and became available for general study in 1990. It has been the darling of chemists, physicists, and material scientists ever since.

A buckyball is the most symmetric molecule possible in three-dimensional space. A perfect sphere, it has 60 carbon atoms arranged like a hollow soccer ball or geodesic dome. In fact, its full name is buckminsterfullerine, in honor of Buckminster Fuller's geodesic dome. The buckyball's atoms are arranged in 12 pentagons and 20 hexagons. Altogether, they measure a mere billionth of a meter in diameter.

▶239 What do bunny balls, fuzzy balls, buckybabies, bucky onions, bucky tubes, and buckygyms have in common?

All are fullerenes, the new form of hollow, all-carbon molecules. Fullerenes have been outclassed celebrity-wise by their most famous member, the buckyball. But chemists have made many more compounds by putting heavy atoms into fullerene cages.

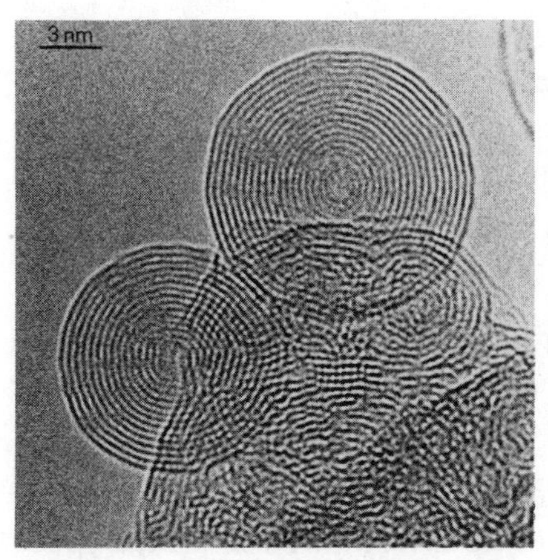

A bucky onion with concentric spheres of carbon atoms. In this transmission electron microscope image, each line is a spherical layer of thousands of carbon atoms.

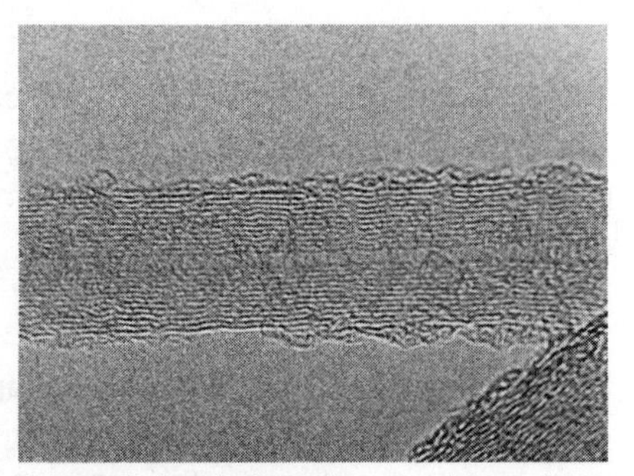

A transmission electron microscope image of concentric layers of carbon atoms that have formed elongated bucky tubes.

A bunny ball looks like a 60-atom sphere with ears. Fuzzy balls have 60 carbon atoms *and* 60 hydrogen atoms. Buckybabies are almost spherical, mostly smaller molecules with 32, 33, 50, 58, and 70 carbon atoms. Onions are concentric shells of carbon, 70 shells deep. And buckygyms are repeating, jungle-gym-like structures made of seven-sided rings of carbon.

▶240 Fullerenes may be fun in the lab, but do they actually exist in the real world? And if so, where?

A. In the soot of candle flames.
B. In interstellar dust.
C. In rocks.

Yes, A, B, C. Blow out a candle and watch buckyballs float through the air. Carbon-60 forms when carbon condenses slowly at a high temperature. So buckyballs are abundant in any sooting flame, candles included.

Buckyballs also may be among the most common and ancient molecules in the universe. They were produced 10 or 20 billion years ago with the first generation of stars. Since they offered the first real surfaces in the universe, other atoms may have clung to them to create matter.

Fullerenes have been discovered in nature in samples of shungite and fulgurite minerals. Shungite is a 600-million-year-old, coal-like rock found only in the small Russian village of Shunga near the Finnish border. Fulgurite is formed when lightning strikes. The carbon content of the fulgurite sample, collected in Colorado, is so low that the fulgurites may come from pine needles carbonized by a lightning bolt.

▶241 What soft crystal becomes harder than diamond when it is squeezed to 70 percent of its volume? And what springs back to its original soft shape when the pressure is released?

You guessed it: a buckyball. It is remarkably resilient. If you hurled one against steel at 17,000 miles per hour (roughly the orbital speed of the space shuttle), it would bounce back.

▶242 Diamonds, candlelight, and stardust are for romantic lovers. But industry likes them too. Why?

When the 60 carbon molecules in a buckyball are crushed under 200,000 atmospheres of pressure at room temperature, they become diamond. Graphite can also be converted into diamond, but at much higher pressures and temperatures.

Buckyballs, which are present in interstellar dust and sooty flames, are ultra-small ball bearings. Because they are perfectly round, edgeless, uncharged, and unbound to other atoms, they spin 100 million times each second. They may become a new kind of molecular lubricant.

▶243 What can almost instantaneously heat a bubble in a liquid until it is as hot as the sun's surface and compress it to pressures of hundreds of atmospheres and then cool it at a rate of more than 1,000 million degrees Centigrade per second—all without heating the bulk of the liquid?

Ultrasound. Sound pitched above human hearing can introduce intense bursts of energy into molecules and cause profound structural and chemical changes. In less than a microsecond, ultrasound can produce temperatures of 5,000°C (9000°F) and pressures like those on the ocean floor. Ultrasound is already used in medicine and industry to treat kidney stones and cartilage trauma, to image fetuses, to make dog whistles and burglar alarms, and for welding and cleaning. "Sonochemistry" is expected to produce new and cheaper catalysts.

▶244 The advertising industry works hard to associate carbonated water with fun and frolic. But this normally harmless substance turned killer near an African crater lake. What happened?

Carbon dioxide bubbled to the surface and formed an invisible cloud. The gas, which is heavier than air, moved at ground level over the countryside around Lake Nyos in the Cameroon. Within hours, it had suffocated 1,746 unsuspecting people and 3,000 cattle.

The carbon dioxide apparently rose from Earth's interior to become dissolved in groundwater over thousands of years. From there, it drained into Lake Nyos, where a warm layer of water floating on top trapped the cold, carbonated water below. The tragedy occurred in 1986 when, for some unknown reason, the layers flipped over, releasing the carbon dioxide. A similar tragedy two years earlier killed 37 people at nearby Lake Monoun. Incidentally, these are the only known cases in which people have been killed by a gas released from a lake.

The same phenomenon can be produced in a mixing bowl by combining a half cup of baking soda and a half cup of vinegar. If a lit match is slowly lowered into the bottom of the bowl, the carbon dioxide there will smother the flame.

▶245 Glass is one of the strongest materials known. Under certain conditions, flaw-free glass is ten times stronger than most common metal alloys. Yet glass can self-destruct. How fast does a glass crack grow?

A. Hundreds of yards per second—at half the speed of sound in glass.
B. Less than one-trillionth of an inch per hour. At that rate, one bond breaks each hour between the silicon and oxygen atoms in the glass.
C. At highly variable speeds, depending on how wet the glass is.
D. All of the above.

D, all three.

A. Glass can seem to shatter instantaneously.
B. In other cases, microscopic cracks take years to grow.

C. Water can make slow cracks move a million times faster. Glass is made of silicon and oxygen atoms bound together; water molecules attack the bonds between the atoms at the root of each crack.

▶246 When Emperor Napoleon III of France invited friends to dine, he gave most of them the usual gold and silver cutlery. His guests of honor, however, ate with knives and forks made of another, much more precious metal. What was it?

Aluminum. The third most common element on Earth (after oxygen and silicon), it is so chemically active that it never appears naturally in its pure form. Small amounts of pure aluminum, however, became available after 1855 for about $100,000 a pound. In 1886, Charles M. Hall in the United States and Paul Heroult in France independently discovered a relatively cheap, electrolytic process for producing pure aluminum. Within months, its price had dropped to 50 cents a pound. Napoleon's elegant aluminum table service had become declassé and was probably relegated to the servants' quarters after that.

▶247 To protect their car radiators down to −30°C (−22°F), many Canadians use antifreeze that is 50 percent ethylene glycol. Why are the body fluids of the gall moth caterpillar about 40 percent glycerol?

To act as antifreeze too. Cars and gall moth caterpillars use the same chemical type of antifreeze: low molecular weight sugar alcohols. Frozen alive with one-fifth of their body weight in antifreeze, the insects survive −38°C (−36.4°F) winters. The caterpillars produce the antifreeze each fall.

Other insects like spiders, ticks, and mites rely on proteins instead of antifreeze to slow the formation of ice crystals and to keep the crystals small. But the proteins can keep the insects safe only to about −15°C.

▶248 Antarctic oceans are filled with perchlike fish that never freeze. What's their brand of antifreeze?

Proteins and glycoproteins, proteins combined with carbohydrate groups. They coat the surface of each tiny ice crystal as it forms inside the blood. The coating prevents the ice crystals from growing big enough to harm the fish.

Relatives of the perch account for 90 percent of the fish in the Antarctic sea. They live for decades in water whose average yearly temperature is below freezing: −1.87°C (28.6°F). Even in summer, ocean temperatures there rarely rise above 2°C (36°F). The fish maintain a year-round supply of antifreeze because their kidneys do not excrete it.

The same proteins that protect Antarctica's fish may also extend the life of donor organs before transplantation. And on a lighter note, ice cream manufacturers hope the protein molecule will keep their products smooth and crystal-free.

▶249 What are "boiling stones," and why do they boil and bubble when heated?

Zeolites, which are highly porous crystals so highly veined with submicroscopic, water-filled channels that they appear to "boil" when they are heated hot enough. In fact, they seem to bubble as the water vaporizes and leaves the crystals. About 40 zeolites exist naturally, and more have been created synthetically. Organized as tetrahedrons, they are composed of silicon, aluminum, and oxygen. Zeolites get their name from the Greek word *zeo*, to boil, and from *lithos*, stone.

Removing the water from zeolite crystals makes them useful as reverse molecular sieves: they trap *small* molecules and let *large* ones escape. Synthetic zeolites are used also as catalysts in petroleum cracking.

▶250 Either while riding on top of a London bus or while sitting before a fireplace in Ghent, Friedrich August Kekulé had one of the world's most famous dreams: a circle of snakes chasing their tails. The snakes were supposed to have revealed to Kekulé the circular structure of benezene, one of the most important discoveries of nineteenth-century organic chemistry.

But were the tails a tale? Kekulé had read earlier in a book about the circular structure of benezene. Who wrote the book? Who is the real discoverer of the circular structure of benzene? What unsung hero was one of the nineteenth-century's leading chemists?

Josef Loschmidt, an obscure Austrian high school teacher of physics and chemistry. According to some scholars, his 1861 book contains the circular structure of benzene in detail. We know that Kekulé read Loschmidt's book a few months after its publication; Kekulé complained that the book was filled with "confusion formulas." Furthermore, when Kekulé published his own, cruder benezene circle four years later, he mentioned Loschmidt's work again.

Luckily, Loschmidt made another great discovery in 1865. This time it was in physics. His calculations for the diameter of an atom and the distance between atoms in a gas are remarkably close to present accepted values. Several physicists recognized Loschmidt's genius and arranged for his appointment to the University of Vienna. There, despite the fact that he had no Ph.D., he became a famous professor and a dean.

▶251 The ABC's—alum, bleach, and city air—have been found guilty of wholesale destruction. What do they do?

Make books self-destruct. Many books today are so brittle that they cannot withstand a single fold. The main culprits are acids, including alum (aluminum potassium sulfate), chlorine bleach, and rosin (an organic acid produced from wood). Paper made before the sixteenth century was acid-free; such paper lasts hundreds of years. From the mid-seventeenth century through the nineteenth century, however, papermakers used increasing amounts of acid. The Library of Congress estimates that about 25 percent of its books are so brittle that they are probably unusable. Happily, this book is printed on acid-free paper.

Although acid-free paper is inexpensive to produce, papermakers must retool their plants. Half the paper produced in Europe is now acid-free, compared to only about 15 percent of U.S. paper.

Heat, humidity, and air pollution also help destroy paper. The sulfur dioxide in heavily polluted air can lower the fold endurance of paper by 15 percent in ten days. For every increase of 6°C (11°F), the life of paper is halved.

▶252 The dregs of a drinking session 5,500 years ago have produced chemical proofs of two of the world's most popular substances: beer and wine. A room in a Sumerian trading post in western Iran housed pottery vessels containing residues of the beverages. The archaeologists' problem is

that the Mesopotamians domesticated barley much earlier, about 10,000 years ago. So which came first: barley beer or barley bread? Did early humans first seek booze or bread?

Archaeologists don't know yet. What they do know is that the room in the Godin Tepe ruins must have witnessed some rather "serious" drinking around the year 3500 B.C.

▶253 Forget the poem about "a jug of wine, a loaf of bread, and thou" and explain why bread is like good wine.

Because both flour and grape juice depend on yeast and improve with age—up to a point, anyway. Add cheese, and you have a meal that depends on microorganisms for all its best features. Storing bread flour for up to one year produces progressively whiter bread, more volume per pound, and finer, softer crumbs. Exposure to air oxidizes the flour and bleaches it. After one year, however, storage begins to degrade the flour. For the past 50 years, oxidizing agents like benzoyl peroxide have been added to new flour to simulate this aging process.

▶254 Mama's little baby loves shortening, shortening,
Mama's little baby loves shortening bread.

By Jacques Wolfe, based on a traditional folk song

Why are some cooking fats called "shortening"?

Shortening shortens the long strands of wheat gluten and makes baked goods softer and lighter. Shortening is any edible, manufactured fat or oil except butter-substitutes like margarine.

Shortening produces a light, tender dough by literally "shortening" or softening the long strands of wheat gluten that make flour tough and elastic. By blocking the interaction between gluten and water, the fat prevents the gluten from reaching full strength. Shortening-rich products are usually mixed with very little water anyway and depend for their structure on the fat's holding the flour particles together. In short, short-stranded gluten is less cohesive and can better incorporate air bubbles and suspend flour and sugar particles.

▶255 **Why is a young frying chicken more tender than an old stewing hen?**

Young animals have fewer cross-links between the layers of their connective tissues. As animals age, they produce more cross-links to hold the protein layers stiff.

▶256 **During the 1960s, each astronaut in the *Mercury, Gemini,* and *Apollo* missions had three tailor-made space suits: one for tests, one for space missions, and one "just in case." The suits were made of advanced materials like Teflon-coated fiberglass, aluminized Nomex, and a variety of rubbers. They were designed to be airtight, to withstand temperatures from −150° to +120°C, and to ward off micrometeoroids. The suits cannot withstand time, however. After only a few decades, the suits are twisted, brittle, and flaking. Which of the three types of space suit is in the worst shape?**

The testing suits. The astronauts trained for weightlessness in swimming pools, and the suits they wore there show chlorine damage. All training space suits in the National Air and Space Museum in Washington, D.C., show deterioration though. Apparently, they were made from "the wrong stuff."

▶257 **Besides their potent perfume, what do rotten eggs, garlic, and onions have in common?**

Sulfur compounds. Crushing or cutting an onion (*Allium cepa* L.) or garlic (*Allium sativum* L.) releases organic molecules featuring sulfur atoms and rare bonding forms.

Thanks to sulfur compounds, the onion is one of the few natural substances that can make a person weep. The compounds also have many therapeutic benefits, however. As antibacterial and antifungal agents, they help onion and garlic plants resist decay underground. In patients, the compounds inhibit blood clotting as effectively as aspirin. Compounds in onions act as antiasthmatics, while others in garlic lower cholesterol and lipids in the blood. Epidemiological evidence suggests that stomach cancer rates are lower when large amounts of garlic, onion, and their relatives are eaten.

Their powerful odor, again traceable to sulfur compounds, has inhibited their popularity as medicine, however. Once eaten, the compounds enter the bloodstream and get exhaled through breath and perspiration. Not surprisingly, some patients object to smelling like a rotten egg.

▶258 The Danish nuclear physicist Niels Bohr escaped capture by the Nazis during World War II by catching a midnight fishing boat to neutral Sweden. He left two gold Nobel Prize medals at home. How did he hide them?

The gold Nobel Prize medal for physics.

A German friend, physicist Max von Laue, had given his Nobel medal to Bohr for safekeeping. As German troops marched through the streets of Copenhagen, the 8-ounce medals were dissolved in a jar of aqua regia. This mixture of nitric and hydrochloric acids ("royal water" in Latin) is one of the few substances that will dissolve gold and platinum. After the war, Bohr precipitated the gold out of the acid and the medals were recast.

▶259 When the Episcopal Church built the Washington National Cathedral during the 1920s, Anglophiles declared that only English Gothic architecture would do. That meant a lead roof, like those on fourteenth-century English cathedrals. But within three years of its completion in 1919, the roof was leaking buckets. Investigation revealed that it was actually slipping and sliding off the top of the building. What caused The Case of the Slipping Roof?

Pure lead is so soft that it can be cut with a fingernail. Its melting point is low, too. On a hot summer day, a roof in Washington, D.C., can reach 80°C (175°F). The English, on the other hand, consider even 28°C (82°F) appallingly warm. Simply put, the roof, which was nearing its melting point, was dripping like molasses. Only when the lead was removed and mixed with 6 percent antimony did it became hard and stiff. Then the cathedral could pack away its pails and puddle-catchers.

▶260 Why would a skier who gets cold on the slopes start rusting away?

To get warm, a skier might use the heat energy released by rusting iron. Exposing a fine iron powder to the air allows oxygen to oxidize the iron. Because the surface area of the powder is large, the reaction proceeds rapidly, and as the iron rusts, it gives off heat. Manufacturers have harnessed this process by packing small amounts of powdered iron in handy packets that, when opened, give off enough heat to keep the skier's hands and feet toasty warm for several hours.

Hand and foot warmers filled with powdered iron.

▶261 How is a hair strand like a space shuttle?

Both have protective coats of hard, light-reflective tiles. Light-colored tiles on the space shuttle prevent heat buildup. Hair fibers are protected by hard, transparent cuticle cells arranged like overlapping roof shingles. Heat, brushing, and harsh chemicals like bleach and hair permanents can dislodge and damage the cuticle cells and reduce light reflection. Damaged hairs snag and tangle; their friction causes static electricity. Silicon-based hair conditioners smooth the hairs and prevent the problem. Sounds like a shampoo commercial, doesn't it?

Astronomy

▸262 **The United States employs 20,000 professional astrologers. How many astronomers work in the United States?**

5,300. Most belong to the American Astronomical Society (AAS), the professional organization of astronomers in the United States.

▸263 **From Hadrian's Wall and the Great Wall of China to the Berlin Wall, people have been building "great walls" for centuries. In 1989, astronomers discovered the greatest wall of all. What is it?**

The Great Wall, a region of galaxies grouped in clusters and superclusters that stretch across the heavens. The largest structure yet discovered in the universe, it is more than a half a billion light years across.

▸264 **There is more to the universe than meets the eye. More gravity acts on and within galaxies than can be explained by the gravitational pull of their visible stars. As a result, most astronomers believe that a form of nonluminous matter, "dark matter," exerts gravitational forces too. What percentage of the universe is supposed to consist of "dark matter"?**

A. 10 percent.
B. 50 percent.
C. 90 percent.

C, 90 percent. Thus, a galaxy may resemble an iceberg: only a tenth of its matter is visible. Actually, dark matter may account for even more than 90 percent of the universe. By some estimates, all the atoms in the stars that are visible now to astronomers represent just 1 percent of the matter in the universe. If so, the universe should contain 100 times more matter than is visible in the stars.

▶265 Assuming that speculation turns out to be fact and not fiction, which of the following is "dark matter" made of?

A. Cold dark matter.
B. Black holes.
C. Dead stars.
D. WIMPS.
E. Hot dark matter.
F. Exotic particles never before observed.

All of the above are possibilities. A. The term "cold dark matter" suggests particles that would move relatively slowly without absorbing or radiating light. B. Black holes are regions of space where gravity is so strong that nothing can come out, not even light. C. Dead stars are stars that have burned up all their hydrogen fuel, cooled off, and become dark. D. WIMPS ("Weakly Interacting, but Massive Particles") include exotica like neutrinos with mass, axions, squarks, photinos, gluino, and so on. E. "Hot dark matter" refers to matter that is so hot that all its electrons have been removed from their atomic nuclei. Stripped of their surrounding electron clouds, the atoms in this "hot matter" cannot absorb or emit radiation and are therefore "dark." They might better be called "hot, transparent matter." F. Astrophysicists have imagined hypothetical particles with names like axions and neutralinos.

▶266 What is the best-mapped planet in the solar system?

Venus, not Earth. Thanks to the robotic spacecraft *Magellan,* more than 98 percent of Venus's surface is mapped, revealing details as small as 125 meters (400 feet) across. Enhanced by computer, images from *Magellan* revealed details the size of a football stadium. Venus's surface is shrouded with 40-mile-thick sulfurous clouds, but *Magellan*'s radar penetrated them. Earth and her twin planet were originally composed of similar matter, but Venus's water evaporated long ago. On the other hand, 71 percent of Earth's surface is covered by water, and large portions of the ocean floor remain unmapped.

Magellan, which began orbiting Venus in 1990, revealed a Venus covered with enormous volcanoes, lava flows, and craters. Most of its volcanoes apparently erupted long ago, but some may still be active. The robot also found traces of plate tectonics, deep trenches like those around the Pacific Ocean, where portions of Earth's outer shell duck under other plates.

▶267 If you're hunting for meteorites, where should you head?

Antarctica. The world's coldest continent is also the world's best meteorite hunting ground. Meteorites elsewhere get plowed under, paved over, and eroded beyond recognition within a few hundred years. But in Antarctica, ice and snow can preserve them for a million years. Meteorites crop up in so-called ablation zones, where mountains bring moving ice masses to a halt and expose them to the wind. Some areas have only meteorites; in others, meteorites sit cheek by jowl with Earth rocks. Meteorites can be distinguished by the black coating that formed when their surface was vaporized as they fell through Earth's atmosphere.

▶268 At 15 billion years of age, blue straggler stars should have cooled off and settled into old age long ago. Instead, they have the energy of rambunctious kids only 1 billion years old. What have blue stragglers discovered?

A. The Fountain of Youth.
B. Antiaging unguents.
C. Hormonal injections.
D. Something else entirely.

D, something else entirely. Blue stragglers are astronomical parasites. In collisions with other stars, they form binary or double star systems and borrow their partner's energy. They may merge with their partner, share its gases, or siphon off some of its mass. The Hubble Space Telescope solved the mystery of the blue stragglers' energy in 1991.

▶269 **On the third morning of space shuttle *Challenger*'s second flight in space, the crew noticed something odd: a pea-sized cavity on the outside of a front window. It looked like a car windshield struck by a pebble. After the flight, the window was removed (cost: $50,000) and analyzed in a scanning electron microscope. Traces of aluminum, carbon, potassium, and titanium dioxide were found. What had struck the *Challenger*?**

A speck of white paint, as big as a grain of salt, traveling at 8,000 miles per hour.

Thousands of tons of rubbish from Earth litter space, including defunct satellites and nuclear reactors, exhaust and paint particles, trash from the Soviet space station *Mir*, and fragments from U.S. weapons tests. Even small bits can be dangerous. Getting hit by an orbiting marble would feel like a 400-pound safe falling on you from the top of a ten-story building. By the year 2000, space ships are expected to face a 50-50 chance of a catastrophic collision. Space walks in low-earth orbits will soon be too risky.

▶270 **At noon on Mercury—the planet closest to our sun—temperatures sizzle at 825 Kelvin (1,025°F). Metals like zinc, tin, and lead would melt into puddles at this temperature. Yet astronomers claim that Mercury has an ice cap. Isn't this as improbable as the proverbial snowball in hell?**

Yes, but it may still be true. Unlike Earth, Mercury's north and south poles never experience a midnight sun in summertime. In fact, the sun never rises much above the horizon at Mercury's north pole. So Mercury has no seasons. The sun's gravitational pull keeps its nearest planet locked into an upright position so that it never tips backward or forward. As a result, its equator always faces the sun directly, and the poles may not warm enough to melt ancient reservoirs of water ice.

▶271 **Twinkle, twinkle, molecular cloud,**
How I wonder what you are,

Up above the world so high,
Like a diamond in the sky.

What is in the molecular clouds of gas and dust that swirl around the Milky Way?

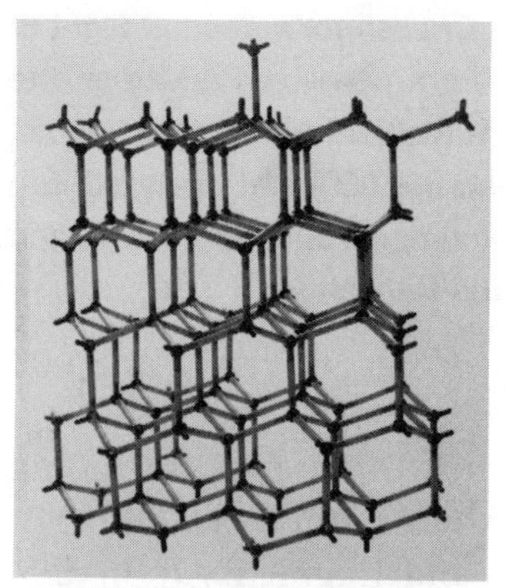

A stick model of a microdiamond. Each carbon atom, represented by a small sphere, is connected to four other carbon atoms in the characteristic pattern of diamonds. Long sticks represent bonds between atoms, and little knobs are atoms on the diamond's surface.

Among other things, diamonds. Not Elizabeth Taylor–sized trinkets, of course, but diamonds nonetheless. Microscopic in size, they measure less than one-millionth of a centimeter in diameter. But there are billions of tons of them, enough to form a jewel the size of a planet. Microdiamonds may constitute 10 to 20 percent of all carbon in interstellar space. They may be free-floaters. Or they may stick to dust grains in the dense molecular clouds that coalesce into stars, planets, and comets. The diamonds may be the remains of stars that died long ago, or they may have formed within the clouds. In any event, microdiamonds have been found in Milky Way clouds and in meteorites that have fallen to Earth.

Astronomers are thrilled, but gemologists may not be. Microdiamonds are so coated with hydrogen atoms that they do not sparkle or twinkle at all.

▶272 **Supernova 1987A was the astronomical event of the twentieth century. It showed astronomers how supernova explosions of stars that have burned up all their hydrogen and helium could have produced virtually all the other elements in the universe. If SN1987A used up all its hydrogen and helium before the explosion, where did it get the energy to explode and eject its newly formed elements into the cosmos?**

From gravitation. The energy could not have come from nuclear energy; that was used up to form the heavier elements, those from carbon to iron. When the core of the star collapsed, the infalling star matter became heated to very high temperatures. According to astrophysicists, the elements heavier than iron were produced in the actual explosion.

▶273 **Where is the worst known greenhouse effect?**

On Venus. The planet is covered by dense, 40-mile-thick, yellow clouds of carbon dioxide topped with sulfuric acid. In addition to the polluted air, the surface temperature on Venus reaches 480°C (900°F). As they say on Earth, though, "It's not so much the temperature that gets you as the humidity." The clouds rain sulfuric acid, although the raindrops evaporate before they hit the ground. As if that weren't unpleasant enough, the atmospheric pressure on Venus is 90 times that on Earth. A day on Venus would feel like forever, or, to be precise, like 243 Earth days.

▶274 **The orderly and predictable motion of the heavenly bodies is often called "the music of the spheres." What kind of music do the planets in our solar system dance to?**

A. Hayden's symphonies.
B. Acid rock.
C. Wagner's ring.
D. Reggae.

It depends on what your idea of chaos is. The orbits of our planets, long thought to be the paradigm of order and predictability, are actually chaotic and hence unpredictable over the long term. Small wobbles in rotations or conflicting gravitational effects from neighbors can cause enormous, exponential changes in the orbits and tilts of planets millions of years in the future. Chaotic wobbles in Mars's orbit have made the tilt of its spin vary by as much as 11 to 49 degrees. As a result, Mars has experienced wild and unpredictable fluctuations in its climate over the eons.

▶275 **Ever since the telescope was invented, the sun's age-old reputation as an unblemished sphere of fire has been going downhill. Now, it turns out that, in addition to spots, flares, prominences and holes, the sun:**

A. Quivers and shakes.
B. Heaves up and down with waves 70 kilometers (45 miles) high.
C. Makes waves that rise and fall every five minutes.

All of the above. Helioseismography, a field of astronomy that began in the mid-1970s, uses solar vibrations to calculate characteristics of the sun's core.

▶276 Hydrogen, the lightest element, is a gas on Earth. Most hydrogen here is bound to oxygen as a component of water. In fact, pure hydrogen molecules are extremely rare on Earth. Since they consist of only two protons, they are so light that they tend to float off into space. Inside Jupiter and Saturn, however, hydrogen is totally different. It is ______________. (Fill in the blank.)

A liquid metal. As condensed-matter physicists enjoy saying, "Just squeeze anything hard enough, and it turns into metal." Saturn and Jupiter are the two heaviest planets in our solar system, Jupiter being the heaviest. As gravity pulls its mass toward the center, enormous pressure builds up on its inner core. Hydrogen atoms there are squeezed closely together. As their electrons move from atom to atom, the material takes on the properties of metals.

Unfortunately for physicists who want to turn hydrogen gas into metal, they work on Earth instead of Saturn or Jupiter. So far, their reach has been bigger than their squeeze.

▶277 Like the dark clouds in the Milky Way, the vast spaces between the stars were long thought to be empty voids. Actually, they form a frozen, chemical-rich brew from which molecules and stars can be made. How many different varieties of interstellar molecules have been identified?

A. 12.
B. 50.
C. 100.
D. 200.

C, 100, although astronomers expect to double that. The molecules exist as gases in enormous clouds that swirl between the stars. Approximately 65 of the molecules are organic and contain carbon.

These include hydrogen cyanide (HCN), formaldehyde (H_2CO), and ammonia (NH_3), the precursors to proteins. Thus, interstellar matter contains many of the molecules necessary for creating life-forms. These life-forming molecules may have been transported to early Earth by comets.

▶278 Aside from moonbeams and moon rocks carried home by astronauts, how much of the moon is on Earth?

So far, more than eight chunks of the moon have fallen to Earth and been discovered. The largest lunar meteorite found so far was on Antarctica in 1989. At 1.5 pounds, the moon rock was ten times bigger than any found before. A NASA astrophysicist picked it up, thinking it looked like an interesting meteorite. But no one knew it was a piece of the moon until it was analyzed later.

▶279 Nuclear fission is the most efficient process on Earth for converting mass into energy. But fusion, the process that makes the sun and other stars burn brightly, is five times more efficient than fission. What energy conversion device is *ten times* more efficient than even the fusion of the stars?

Falling into a black hole. Black holes are thought to be collapsed bodies that exert such tremendous gravitational forces that nothing, not even light, can escape them. As nearby stars and gases swirl into the black hole, friction heats their atoms even further. Soon their electrons are stripped off, and the atoms fall apart, creating an electrically charged gas called plasma. Then the black hole's gravitational forces squeeze the magnetic fields within the plasma, heating it to unimaginable temperatures. As more and more matter falls into the black hole, the region around the black hole becomes intensely bright. According to speculation, a quasar may be created. Quasars are the brightest regions in the universe.

▶280 How many black holes have been discovered?

None. Zero. Zilch, whichever you prefer. Black holes remain an attractive, but unconfirmed theory. Even if the theory is proven correct, however, it would be most unwise to go exploring for one. A black hole is the concentration of a mass with such a strong gravitational field that not even the light from its own surface can escape. Hence its name. Black holes are thought to be created at the end of a massive star's life when there is so much matter in one place that no force can keep it from collapsing on itself. It was once thought that nothing emerges from a black hole, but Stephen Hawking showed in 1970 that it is possible for black holes to emit radiation. Through Hawking radiation, small black holes could "evaporate" with time.

▶281 **Con artists can always find a buyer for a perpetual motion machine. So what energy conversion process is theoretically as efficient as these imaginary devices? By converting 100 percent of matter into energy, this process would be even more efficient than fission, fusion, or black holes.**

The annihilation of matter by antimatter, because it liberates all the energy contained in mass. Unfortunately, physicists have succeeded in making only a few antiparticles at a time. One of the mysteries of the universe, in fact, is why the cosmos is composed of matter, rather than equal quantities of matter and antimatter.

▶282 **What can be the most dangerous time in the life of a space vehicle?**

A. On the ground moving from factory to launch pad.
B. During a launch.
C. During space travel.
D. At reentry.

A. Earthbound travel can involve more shocks and vibrations over more time than a normal launch or flight. Moreover, though space vehicles are designed for space, little consideration may have been given to their needs during ground transportation.

▶283 **A captain in the Argentinean air force was making an aerial survey of the Pampas in 1989 in preparation for an international model-airplane contest. Flying along at an altitude of 900 meters (3,000 feet), he spied a long and irregular crater gouged across the cultivated fields below. Returning the**

next day in a jet trainer, he flew much higher than the day before. Suddenly, he saw a series of ten long, thin craters stretching out in a line for more than 30 kilometers (18 miles). What had the aviating astronomy buff discovered?

Evidence of a rare kind of impact between an asteroid and planet Earth. Perhaps as recently as 2,000 years ago, an asteroid grazed Earth at a nearly horizontal angle. Similar impacts have been observed on the moon, Mars, and Venus and simulated in ballistics laboratories. But none had ever been observed on Earth.

The asteroid, approximately 150 meters across, was traveling at an estimated speed of 55,000 miles per hour. Approaching at a 15-degree angle, it burst into fragments. They struck Earth at a shallow angle, forming long craters as they hit. Inside the depressions were found asteroid rocks and glassy fragments, formed when heat from the explosion melted surrounding rock. Not a bad prize for helping to organize a model-airplane contest.

What objects were once called "vermin of the skies"?

A. Asteroids.
B. Meteorites.
C. Man-made trash.
D. Intergalactic dust.

A, asteroids. These small rocky objects, sometimes called "planetoids," were once regarded as the lowest of the low, celestially speaking. Now they are considered to be the remnants of planets that failed to form between Mars and Jupiter. They were prevented from merging together by the pull of Jupiter's enormous gravitational force. The largest asteroid is 933 kilometers (580 miles) in diameter, and about 1,000 asteroids measure more than 30 kilometers (19 miles) across. Nevertheless, most asteroids are quite small. Astronomers estimate that there are approximately 1 million between 1 kilometer and 30 kilometers across (half a mile and 20 miles). Asteroids are probably the ultimate source of most meteorites that fall on Earth.

▶285 Although Mercury is much smaller than Earth, it contains far more iron. In fact, it has more iron than any other planet in our solar sys-

tem; three-quarters of Mercury consists of the element. As a rule, the more iron a planet has, the denser it is. So which is the densest planet in our solar system?

Earth, *not* Mercury. Earth's density is 5.52 grams per cubic centimeter, compared to Mercury's at 5.44 grams per cubic centimeter. This is because Earth is more than twice as big across as Mercury. Thus, the sheer weight of Earth compresses its volume and makes it denser. Without this compression effect, Mercury's density would be greater: It would be 5.3 grams per cubic centimeter, compared to Earth's at a mere 4.4.

▶286 **What and where is the world's largest optical telescope?**

The 10-meter W. M. Keck telescope, opened on top of Mauna Kea in Hawaii in 1991. The telescope's light collector is four times more powerful than that on Palomar Mountain in California. Its eye is composed of 36 hexagonal mirrors, each measuring 1.8 meters across. Hundreds of microcomputers, sensors, and mechanical devices align them. With 76 square meters of collecting area, the Keck telescope should be able to probe deep into the universe to observe galaxies as they existed 12 billion years ago.

A second, identical telescope is to be opened next door on the same mountain peak so that the two can operate in tandem.

▶287 **The 200-inch telescope at Palomar, California, is the third largest telescope in the world. Nevertheless, over the past 45 years, it has effectively "shrunk" from 5.1 meters to 3.5 meters. It now operates at roughly 39 percent of its original efficiency. What "shrank" the scope?**

Light pollution. Palomar Mountain is disturbingly close to the night lights of San Diego. Palomar's night sky is now twice as bright as it would be without San Diego. Thus, the telescope is considerably less efficient than it used to be. What causes the problem? Advertising, safety, and athletic field lights illuminate skies as well as ground. Astronomers want outdoor fixtures shielded so that light shines only where it is needed. One percent of the electricity in the United States is beamed at the sky. Mercury, incandescent, and high-pressure sodium lamps particularly annoy astronomers. Like penny-pinchers, they prefer low-pressure sodium lamps. Besides being inexpensive to operate, sodium lamps are easy to filter out because they concentrate their light in a narrow spectral region.

▶288 How old is the Milky Way?

A. 12 to 18 billion years old.
B. 9 to 15 billion years old.
C. 7 to 11 billion years old.
D. 13 billion years old.

It depends on how you figure it. Judging by the color and luminosity of globular clusters of stars, the Milky Way would be 12 to 18 billion years old. Based on radioactive isotopes in stony meteorites, the galaxy is younger: between 9 billion and 15 billion years old. Based on counts of white dwarfs, which are dying low-mass stars, the Milky Way is only 7 to 11 billion years old. The most favored age, however, is about 13 billion years.

▶289 What is the most volcanically explosive part of our solar system?

Io, the innermost moon of Jupiter. Unless some of Venus's volcanoes prove to be active, Io and Earth are the only two objects in the solar system with active volcanoes. And Io outclasses Earth's pyrotechnics by a long shot.

The surface of Io is coated with sulfur in the form of sulfur dioxide and hydrogen sulfide. Since sulfur dioxide reacts with water to make sulfurous acid, and hydrogen sulfide makes rotten eggs smell rotten, Io is not a good vacation stop.

Io is also pocked by active volcanoes. When *Voyager* spacecrafts flew by Io, nine of its volcanoes had plumes 300 kilometers (200 miles) tall. The size of Earth's Moon, Io can throw more than 10 trillion kilograms of matter into its atmosphere each year. Thus, any evidence of craters is rapidly covered up. Jupiter, the largest planet in our solar system, exerts enormous gravitational pull on Io's rocky surface. Heaving with tides, Io bulges back and forth, generating in the process enormous amounts of heat in its interior. Hot gases under pressure explode through fissures in Io's crust.

The volcanoes on Earth and Io spew rock. Neptune's moon Triton, on the other hand, has so-called geysers that shoot nitrogen several kilometers into the atmosphere.

▶290 What part of our solar system guarantees space travelers the coldest welcome?

Triton, Neptune's largest moon. At 37 Kelvin (−236°C or −393°F), Triton is the coldest place in the solar system. Nitrogen, the main component of Earth's air, is a solid on Triton's surface. Lakes of pink ice made of frozen nitrogen stretch across its southern hemisphere. The pink tint comes from the organic compounds formed when radiation bombards methane.

Triton's atmospheric pressure is remarkably thin: ten-millionths that of Earth at sea level. Yet some atmosphere does exist. Nitrogen mixed with a trace of methane extends 800 kilometers (500 miles) up from the moon's surface. And Triton probably has nitrogen eruptions similar to geysers.

▶291 The influence of the moon on Earth is supposed to be lunatic. Even the word says so. But ancient traditions can be luny too. Is this one?

Yes. Earth's orbit is slightly chaotic. But fortunately for us, the moon is a steadying influence. Otherwise we would tip back and forth chaotically, and our seasons and climate would be, dare we say it?—sheer lunacy. Life-forms need a dependable climate. So a steadying moon may be the key to the development of life on other planets. Thus, when trying to find life elsewhere in the universe, try looking for planets with moon-sized satellites.

▶292 Besides the sun, where are neutrinos known to originate?

In Supernova 1987A. Neutrinos, which have no mass or charge, are extraordinarily hard to detect. Supernova 1987A is the only cosmic source of neutrinos detected so far. In a 13-second burst a few hours before the supernova's light became visible to astronomers, two neutrino detectors counted a total of 19 neutrinos. On a typical day, the heavily shielded devices buried in salt and lead mines register a single neutrino.

▶293 **Commemorating Christopher Columbus's discovery of the New World, astronomers in 1991 inaugurated a ten-year, $100 million microwave survey. What are they trying to locate?**

A. Signals from intelligent, extraterrestrial life.
B. A new class of pulsating radio stars.
C. Variations in microwave radiation at the edges of the universe.

A. By using microwaves, they avoid background noise from our galaxy; stars emit less radiation in those wavelengths. Critics of the program quote the late nuclear physicist Enrico Fermi, "If extraterrestrial beings exist, why aren't they already here?" According to the odds, Fermi thought in the 1950s, some of the extraterrestrial civilizations would have to be more technologically advanced than Earth and would probably have visited us by now. Stay tuned, though, for a broadcast from any one of a thousand Milky Way stars within 80 light-years of Earth.

Physics

▸294 **Intergalactic space is sometimes considered the coldest natural place in the universe, but it's warm compared to the coldest spot on Earth. Where on Earth is The Biggest Chill of all?**

A. Polus Nedostupnosti, Antarctica.
B. Bayreuth, Germany.
C. Lancaster, England.
D. Helsinki, Finland.
E. Boulder, Colorado.

D, in a physics laboratory at the Helsinki University of Technology in Finland. Physicists there achieved the lowest temperature that has ever existed—by bringing silver and copper nuclei to two billionths of a degree Kelvin above absolute zero. Absolute zero is zero degrees Kelvin or −459.67°F.

The Finnish physicists cooled atomic nuclei of copper and silver to set the record in 1989. They did it by aligning the nuclei and bringing them almost to a stop.

While the public may be intrigued with setting records, most physicists are more interested in the bizarre properties of matter at such low temperatures. To study these properties, other physicists have cooled electrons and individual atoms to temperatures within one-millionth of a degree of absolute zero. The other earth-bound locations listed in the question have produced different record low temperatures too. But for sheer, unadulterated cold, the Finns and their nuclei have everyone else beat.

In comparison, intergalactic space is positively toasty at approximately 2.7 degrees Kelvin. The total amount of energy there is extremely low because it contains so few atoms. But some of the gases in space may be extremely hot (in the millions of degrees!), so the amount of energy per atom may be very high. In Helsinki and other cold temperature laboratories, the amount of energy per atom is exceedingly small.

▶295 Several groups have refrigerated matter to near-record cold temperatures. But only one group of researchers used relatively inexpensive, tabletop equipment to come within a millionth of a degree Kelvin. How did it work?

By using lasers to hit individual atoms head-on, bringing them to a dead stop. Without motion, there was no heat—or virtually no heat, at any rate.

A small glass container filled with a gas of metal cesium atoms was pumped out. About 100 million atoms remained, bouncing around and colliding with each other and the container. Then photons of light from six infrared lasers were beamed on the atoms from different angles. As each atom was struck head-on, it lost some of its momentum. Hit over and over again, the atoms gradually slowed to a virtual stop. Without motion, there was no heat. Eventually, the temperature in the container measured almost absolute zero, −273.16°C or −459.69°F. The experiment was designed and carried out with about $10,000 worth of equipment at the University of Colorado at Boulder. Enthusiastic news reports whittled the price down to a few hundred dollars.

▶296 The princess and the pea have met their match. What kind of experiment could be so delicate that one fly doing one push-up could wreck the results?

Low-temperature experiments near absolute zero. An exercising fly generates too much heat. Even cosmic rays can deliver about ten-trillionths of a watt. Dropping one ant 1 centimeter onto a table generates as much heat as seeps into a typical low-temperature experiment over one year.

▶297 **Big science, huge machines, and elaborate experiments keep ever-growing teams of high-energy physicists busy accelerating ever-smaller subatomic particles at higher and higher speeds. What's the record number of authors for a single physics article?**

407 physicists. The alphabetical list of authors covers 1⅓ pages of the prestigious journal, *Physical Review Letters*—and that's just using their surnames and first initials. The article itself was a three-page letter. The report, on an experiment in high-energy physics at Stanford University, appeared April 26, 1993. It had the gripping title, "First Measurement of the Left-Right Cross Section Asymmetry in Z Boson Production by e^+ e^- Collisions."

At least they succeeded in measuring something. The earlier record for most authors was an article by 315 high-energy physicists. They had labored mightily on the Tevatron, the world's most powerful particle accelerator, but failed to find the evidence they were searching for. Their article appeared December 14, 1992, in *Physical Review Letters*.

▶298 **String theory deals with objects, called strings, that are unimaginably smaller than atoms, nuclei, quarks, and leptons. If you could do an experiment to test string theory with an accelerator like those used with atomic particles, how big would the accelerator have to be?**

The size of a galaxy. If they exist, strings are only 10^{-34} meters long. Strings are much smaller even than nuclei. In fact, they are even smaller compared to a nucleus than a nucleus is compared to a table, for example.

To go from the size of a table to the size of an atom is to move ten powers of ten smaller (10^{-10}). To move from the size of an atom to nuclei is to go five more powers of ten (10^{-5}) smaller. So far, the move from a table to a nucleus has required a scale-down of 10^{-15}. To get to strings is an even bigger jump down to 10^{-34}. So don't hold your breath, waiting for confirmation of string theory—at least not with an accelerator. Physicists are working hard to find consequences of string theory that can be tested indirectly with the accelerators of today or the near future.

▶299 For 200 years, scientists believed that every solid had to be either glassy or crystalline. Then in 1984, a third type of solid matter was discovered. What is it?

Quasicrystals, which are somewhere between crystalline and glassy. The atoms of a glassy solid are frozen into a completely random arrangement. The atoms of a crystal are organized in a regular pattern that can be repeated over and over again like three-dimensional wallpaper. The atoms of a quasicrystal are arranged in a rather symmetrical pattern, but when the pattern is repeated, some parts look different from others. Quasicrystals do not make three-dimensional wallpaper patterns.

This quasicrystal is made of 12 symmetrical pie shapes. When the pattern is rotated one-twelfth of the way around, it still looks the same. But when it repeats off to the sides, it looks different.

Quasicrystals were first discovered in an alloy of aluminum and manganese made in 1984. Previously, in 1974, a mathematician named Roger Penrose developed a simple set of rules for covering a plane with a few tile shapes without ever repeating the pattern precisely. Since then, many other metallic alloys and tiling patterns have been discovered that may also be quasicrystals.

▶300 Atoms of matter can be broken down into incredibly tiny particles. How many different kinds of particles do physicists recognize?

A. 24.
B. 200.

A and B. More than 200 different kinds of particles have been identified in the debris from atomic collisions in physics laboratories. These "elementary" particles include photons, protons, neutrons, electrons, neutrinos, mu neutrinos, muons, pions, K mesons, pions, etas, sigmas,

Z and W particles, cascades, lambdas, sigmas, omegas, K and anti-K particles, bottom mesons, D particles, taus, and tau neutrinos.

On the other hand, physicists also believe that most of these particles are made of just six kinds of quarks. The quarks are bound so tightly together that it is impossible to see an individual quark, however. They apparently exist only in groups of twos and threes. Some of the particles, such as photons, electrons, and neutrinos, are indivisible and are not made of quarks. Thus, they appear on both lists.

▶301 When two of the world's smallest atoms get together, they make a cluster the size of a biological macromolecule that has hundreds of atoms. What are the atoms in this giant, two-atom cluster?

Two atoms of helium, the element with the second smallest atoms. Physicists wondered for 65 years whether two helium atoms would ever be attracted to one another enough to form a relationship of any sort. They got their answer in 1993. Two helium atoms will combine as a large but fragile cluster called a dimer—but only under extraordinarily frigid conditions a tad above absolute zero.

The most amazing fact is that two atoms of helium can be made to associate with each other at all. They are so lightweight and inert that they rarely bond with anything. Even as a frigid, fragile cluster, they are held together by the weakest bond ever measured (the result of their electrons being unevenly distributed around their nuclei). The bond is far too weak and fragile for the cluster to be considered a molecule. And the helium atoms remain standoffish: a frosty 55 angstroms apart instead of the usual two or three. Hence the cluster's enormous size. It's hardly what you'd call an intimate relationship.

▶302 Why would a nuclear physicist want to raise 1,500 ingots of lead from a Roman freighter that sank off Sardinia more than 2,000 years ago?

Lead is the best material for shielding delicate experiments from the high-energy cosmic rays that rain down through the atmosphere. But lead itself contains small amounts of radioactive material too. Because the half-life of radioactive lead-210 is 22 years, lead excavated 2,000 years ago should have lost almost all its radioactive isotopes by now. It should be even better than the old cannonball lead that physicists sometimes use as shielding.

▶303 **Two physicists won Nobel Prizes for work they did as unpaid volunteers. Who were they, and why weren't they paid?**

Marie Curie and Maria Goeppert Mayer, because they were women.

Marie Curie won two Nobel Prizes for studies of radioactivity. It wasn't until 1904, the year after she received the Nobel Prize in physics, that she secured her first university job: as her husband's lab assistant.

Maria Goeppert Mayer won the Nobel Prize in physics in 1963 for discovering the shell model of the atomic nucleus. It explains how protons and neutrons orbit around inside the nucleus. Mayer worked for 30 years before she received a paycheck from a university.

▶304 **One of the truisms of high school physics is that like-charged particles repel one another. When is there a net *attraction* between two electrons?**

In superconductivity, in which conductors completely lose their resistance and currents flow without cease. In superconductive material, electrons pair off. In some obscure way, this process makes the electrical resistance of the material disappear. A key to this remarkable phenomenon is the fact that if two objects are attracted to a third object, the first two are indirectly attracted to each other too. And if you thought you'd get a clearer explanation than this, you should have heard physicists discussing the possible explanations in 1987 when superconductivity was discovered in ceramic materials. They still have not figured it out.

▶305 **The discovery of high-Tc superconductivity in ceramics in 1987 was supposed to revolutionize modern technology. For the first time, it would be technologically simple to conduct electrical currents with virtually no friction. A technological millennium hovered on the horizon. So far, what commercial applications of high-Tc technology have appeared?**

A. Magnetic trains.
B. Magnetic resonance imaging scanners.
C. Super-efficient telecommunications.
D. New devices to measure magnetic fields.

Only D. None of the others has proved feasible. So far, it has proved impossible to sustain a large current above approximately 125 Kelvin (−234°F). It was the possibility of achieving superconductivity at high temperatures that generated so much excitement in the beginning.

▶306 **Of all the fundamental forces of nature, which is calculated mathematically with the *least* precision?**

Gravity, which, ironically, is also the force that scientists have known about the longest. The force of gravity is known to only four decimal places. All other fundamental constants are known to at least six. A more precise figure would help astronomers calculate more accurately the masses of the Sun, Earth, and other planets; the probability of the universe's ending in a "big crunch" or expanding forever; the brightness of stars; and other problems. If experiments to refine the value of gravity were done in space, interference from Earth's gravity and vibrations would be eliminated.

▶307 **Some containers are as famous as their contents. What are:**

A. Leyden jars?
B. Magnetic bottles?
C. Klein bottles?
D. Paul and Penning traps?

A. The Leyden jar, invented in 1745 by Pieter van Musschenbroeck in the Dutch city of Leyden, was the first cheap, easy-to-use source of electric sparks. A glass bottle was lined with metal foil inside and out and filled with water and an electrified wire. Anyone who touched the wire received a severe shock.

B. A magnetic bottle contains plasma, which is a gas of electrified particles so hot that it melts any conventional vessel. On the sun, for example, plasma reaches a million degrees C (2 million degrees F). As a result, physicists confine plasma inside magnetic fields; they keep the plasma superhot by squirting in neutral particle beams, changing electromagnetic fields, or microwaves.

C. A Klein bottle is an imaginary mathematical construct that has no inside or outside; that is, its inside is the same as its outside. If cut in two, it becomes two Möbius strips, which are also one-sided constructions with neither fronts nor backs. This weird topological jar is named for the German mathematician Felix Klein.

D. A Paul trap is an arrangement of electric and magnetic fields to hold a few atoms for study over many months. Its cousin is the Penning trap. Physicist Wolfgang Paul of West Germany shared the Nobel Prize in 1989 for developing his trap.

▶308 What was so hot about cold fusion?

Nothing, as it turned out. In 1989, two chemists, Stanley Pons and Martin Fleischmann, of the University of Utah held a press conference to announce a spectacular claim: They said they re-created the fusion of the sun in a simple battery jar at room temperature. More energy had come out of the jar than had gone into it, they declared. It was a dream come true for an energy-dependent society. In the flurry of excitement that followed, several other laboratories confirmed their experiment. Since then, however, no one else has been able to. So far, only the stars and thermonuclear hydrogen bombs have been able to fuse nuclei together to create enormous amounts of energy.

▶309 Imagine making portraits of:

— One atom at a time.
— Individual chemical bonds.
— Chemical bonds anchoring individual atoms to bulk material.
— A few atoms placed on a surface. Within minutes, the atoms nestle down into the surface.
— Surface electrons forming strange and complex patterns that change as surface temperatures vary.

Are these portraits:

A. Sci-fi?
B. ESP?
C. STM?
D. STP?

C, STM, short for scanning tunneling microscope. STMs magnify matter a hundred million times and image one atom at a time. The "portrait" is actually a computer-generated topographical "map" of surfaces. An STM operates like a stylus reading braille. The tip of the stylus, as small as one atom at the end, moves over the surface of a bit of matter to reveal its topography. The tip may get within five-atom-lengths of the surface—so close that its electron clouds mingle with those of the surface atoms. An STM still cannot reveal what kind of atom is pictured. But it can show how the surface electrons form strange and complex patterns.

A magnified image of the tungsten tip of a scanning tunneling microscope (STM).

▶310 The Hope Diamond weighs 44.5 carats. How many molecules does it contain?

One. Every diamond, no matter how gigantic, is a single molecule.

Every electron in an atom forms a cloud that surrounds the atom like a fog. Every element on Earth has a different way of distributing these electron clouds around its atoms. In a diamond, the electron clouds spread out and surround all the atoms and hold the diamond rigid. As a result, diamond is the hardest naturally occurring substance known.

▶311 Uranium is the heaviest atom in nature because it has the most protons and neutrons in its nucleus. Because its nucleus is so tightly packed with positive charges, uranium is highly unstable and apt to decay radioactively. Is uranium also the largest naturally occurring atom?

No. The atoms of 26 elements are as large or larger than those of uranium. Sixteen atoms are bigger than uranium: barium, cerium, cesium, europium, gadolinium, lanthanum, neodymium, praseodymium, promethium, rubidium, samarium, scandium, sodium, strontium, thallium, and yttrium.

The size of an atom is determined by its electrons, not by its nucleus. The sheer number of electrons is rather umimportant. The vital question is whether the atom's outer electrons are tightly or loosely bound to the nucleus by their opposing charges.

Some heavy atoms, like uranium, can pack many electrons into a small space. But other atoms have loosely bound outer electrons, and they occupy more space. Cesium is the largest atom. It has only 58 protons, so it is much lighter than uranium with 92. But cesium's outer electrons are very loosely bound to the nucleus.

▶312 Even when it is almost as cold as absolute zero, this substance remains a liquid, ignoring the laws of friction as it flows. What is it?

Liquid helium, which exists in two isotopes: ^{4}Helium, the common isotope, and ^{3}Helium, a rare isotope. At temperatures approaching absolute zero, they become superfluid superconductors. Helium atoms are so light that they interact weakly with one another. As a result, they cannot be kept stationary long enough to form a solid, even at absolute zero.

▶313 Does any element remain a gas at absolute zero?

Atomic hydrogen. Normally, every element is supposed to become a solid at absolute zero, with the exception of helium, which becomes a liquid. But current theory suggests that atoms of hydrogen—as opposed to molecules of hydrogen—could remain a gas at absolute zero too. The electron spins would have to be aligned in the same direction by a strong magnetic field, however, to keep the atoms from forming hydrogen molecules.

▶314 A leaping dancer who appears to "float" through the air is defying the laws of physics. Ironically, the dancer achieves the effect by suddenly pulling himself *down* so that he cannot go as high as he otherwise would. What is going on?

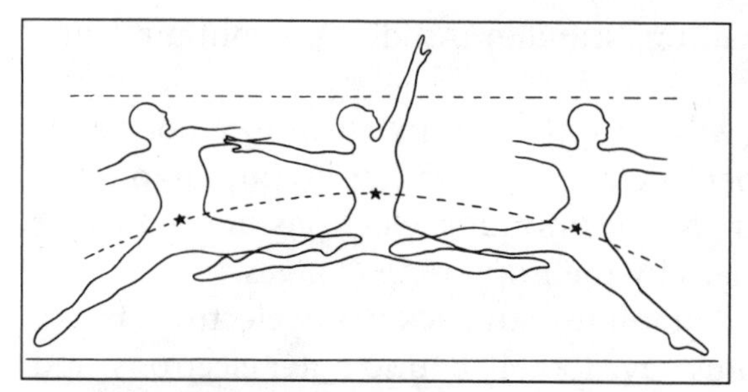

The trajectory of a leaping ballet dancer.

Like any other projectile, a leaping body follows a normal trajectory through the air. First, the body rises, it reaches an apex, and then it falls. Toward the peak, however, our leaping dancer raises his (or her) arms and legs to stop himself from going up any farther. When he thrusts his legs and arms up, the movement pushes his torso down in reaction. As a result, his head—which is what the audience looks at to gauge the height of his jump—stays at roughly the same height for most of the jump. Thus, he appears to hover longer at the "peak" of his leap.

Actually, the trick satisfies the laws of physics as well as the audience. According to natural laws, his center of gravity must move in a arc-like trajectory. At the beginning of his leap, his center of gravity is somewhere in his abdomen. When he lifts his arms and legs, his center of gravity moves up to his stomach. As his limbs fall back down, his center of gravity falls back to his abdomen.

▶315 Solids, liquids, gases, and plasma—the four states of matter—may be joined by a fifth. What would it be?

Clusters, according to some physicists. Tiny groups of atoms, clusters range in size from two to several hundred identical atoms. Surprisingly, they have very different properties from larger groups of the same atoms. The most famous clusters are buckyballs and the other fullerenes.

As the clusters add and subtract atoms, they quickly change states between gases, liquids, and solids. As they change states,their properties change too. For example, their ability to switch back and forth between transparency and opacity makes them prime candidates to power optical computers.

Clusters are formed at temperatures hotter than the surface of most stars. At tens of thousands of degrees, lasers vaporize the surface atoms off solid metals. As the vapor cools, atomic clusters of various sizes form.

Physicists, chemists, engineers, biochemists, and others have jumped on the cluster bandwagon. They use different lingo, though. Engineers say "nanostructures," implying a device. Chemists and physicists talk about "clusters," meaning giant molecules. It does not matter which word they use. Clusters by any other name are just as ultrasmall.

▶316 **When is gold the color of midnight blue?**

When the gold is deposited in the form of clusters on a surface. When a few atoms form clusters, their properties often differ from their solids. Pathek, a Swiss manufacturer, coated the face of an expensive watch with gold clusters. Its color was dark blue.

Clusters, which are easy to magnetize, are also used to produce high-density videotape.

▶317 **The Foehn of Switzerland, the kachchan of Ceylon, and the berg wind of South Africa are foreign versions of what excitingly hot, high-flying phenomenon?**

The Chinook and the Santa Ana.

▶318 **According to legend, the Chinook and the Santa Ana drive animals and people mad. They seem to disobey the fundamental laws of nature. Although they are hot, dry currents of air, they race *down* off *cold* mountaintops at speeds approaching 80 miles an hour. Warm air is supposed to rise, but these winds are often 50°F warmer than the surrounding air. Are they also driving physicists mad by disobeying nature's laws?**

Of course not. The winds are dry because they gave up all their moisture on the ocean side of the mountains. Starting out at the mountaintop, they are cold. As they descend, they get progressively hotter. But they are falling so fast that they cannot transfer their heat to anything else.

▶319 How do you make copper and silver, which are normally soft, as hard as iron? And how do you make brittle ceramics change shape under pressure instead of break?

You make them into an ultrafine froth like soap bubbles. Reducing the graininess of a solid can radically change its properties. First, in a vacuum, beam a laser onto metal to vaporize it into individual atoms. They condense and form a powder a few atoms deep on a rod. When the powder is squeezed hard under high pressure, it forms solid grains a few billionths of a meter across. Such small grains of copper are five times harder than conventional copper. With the spaces between their grains also reduced to a few billionths of a meter, ceramics conduct electricity more efficiently.

▶320 Fogbows, mistbows, cloudbows, dewbows, and rainbows can be seen when you stand with your back to a low-lying sun and face toward a rain or water shower. What makes one 'bow different from another 'bow?

The size of its water droplets. Fogbow drops are less than 0.05 millimeter across. Mistbow and cloudbow drops are smaller than 0.06 millimeter. Dewdrops after sunrise are large enough for all the colors of the rainbow to appear.

All 'bows are caused by light entering a waterdrop, striking the far sidewall of the drop, and then being reflected back out again. The angle of reflection determines the color we see. Large drops more than 1 millimeter in diameter reflect red, green, and violet, but little blue. The smaller the droplet, the weaker the red.

Even moonlight can generate a 'bow. Depending on the size of their water drops, moonbows can be as colorful as rainbows. But most appear colorless because the human eye cannot see much color in dim light.

▶321 Quantum has become a trendy word meaning a big jump or great advance. What's wrong?

A quantum is a set amount of energy, a specific amount. But its size is incredibly small.

▶322 What is the world's teeny-tiniest battery, so small that 100 could fit into the red blood cell of a human?

A 500,000-atom battery composed of a graphite surface topped with two supersmall pillars, one made of copper atoms and the other of silver atoms. The battery generates one-fiftieth of a volt during its 45-minute lifetime. Before it dies, it dissolves about 75,000 copper atoms and coats them two atoms thick over the silver pillar. The device generates an electric field, which may enable researchers to study how proteins in muscle orient in an electric field and how at the atomic level metals corrode. A scanning tunneling microscope was used to build the device.

The cell is an example of nanotechnology, which uses building blocks so ultrasmall that they are measured in nanometers, or billionths of a meter. Matter built of such infinitesimal particles sometimes has totally different properties than familiar substances made of larger particles. Nanostructured ceramics, for example, can be stretched like chewing gum at 1,600°C (3,000°F) and are tougher and stronger than more coarsely grained ceramics. The copper and silver pillars in the battery were incredibly small: between 2 and 5 nanometers high.

▶323 A politician with a spark of genius used two teams of horses, 50 men, and a ball in three famous demonstrations important to atomic theory, electricity, and steam engines. Who was he and what did he do?

Otto von Guericke, the mayor of Magdeburg, Germany, invented an air pump in 1654. To show Emperor Ferdinand III the power of air pressure, he put two halves of a hollow brass ball together to form a sphere and pumped the air out. Outside air pressure pushed the two halves together so powerfully that 16 horses could not separate them. When a valve opened and air entered the sphere, the halves fell apart.

Von Guericke also invented the first machine that made electric sparks; rubbing his rotating sulfur ball produced static electricity. Renaissance air and vacuum experiments helped establish that matter is made of tiny particles called atoms.

The mayor's early steam engine consisted of a piston in a cylinder. After creating a partial vacuum below the piston, he challenged 50 men to keep the piston from moving into the cylinder. They failed.

▶324 **What do the following have in common when mixed with a drop of water and viewed through a microscope: 100-year-old pollen grains and powders made of pulverized pieces of the Sphinx, London soot, window glass, fossilized wood, volcanic rock, meteorites, arsenic, and nickel?**

Brownian motion. The particles quiver and move around incessantly. They are under continuous bombardment from the water molecules, which have enormous amounts of energy even though they are a million times smaller than Brown's pollen grains and powders. Robert Brown, a Scottish soldier and explorer turned botanist, discovered the phenomenon in 1827. At first he thought the energy came from the pollen, so he redid his experiment with many different powdered substances, including those inquired about in the question.

▶325 **Picks and shovels have been joined by high technology on archaeological, anthropological, and geological digs. To date their samples, collectors use a spectacular array of scientific methods. And for the best results, they use several methods and compare the results. Match the technique with the kinds of objects they analyze.**

Technique	*Object of Analysis*
1. Carbon-14 radioisotope.	**A. Ocean sediments.**
2. Thermoluminescence.	**B. Clay pottery and soil.**
3. Uranium thorium radioisotopes.	**C. Coral.**
4. Dendrochronology.	**D. Small fossils, e.g., shells, teeth, and eggshells.**
5. Amino Acid Racemization (AAR).	**E. Shroud of Turin.**
6. Paleomagnetism.	**F. Rocks.**
7. Fission tracks.	**G. Mica.**
8. Rubidium strontium radioisotopes.	**H. Trees.**

1–E. Used with an accelerator mass spectrometer, it is especially useful for analyzing small samples like grain and seeds.

2–B. Energy builds up in the clay from the radioactive decay of trace elements in the clay. When the clay is heated, the energy is released all at once as light. This technique measures how much time has elapsed since the material was last heated.

3–C. The relative amount of radioactive isotopes in an object decreases at a regular rate with time, providing a way to measure the age of the object.

4–H. By analyzing tree ring growth.

5–D. AAR analyzes the rate at which amino acids convert from being one-handed to being a mixture of left- and right-handed molecules (see the section entitled "Chemistry").

6–A. Traces of magnetism in the sediments record the earth's magnetic field at the time that the sediments were formed.

7–G. Uranium in the material undergoes radioactive decay, leaving tracks.

8–F. See 3–C.

Mathematics and Computers

▸326 **What is the world's largest single employer of mathematicians with Ph.D.s?**

The National Security Agency. Modern cryptography has become inseparable from abstract mathematics.

▸327 **When Evelyn Marie Adams won the New Jersey Lottery twice in four months in 1986, experts said she had beaten odds of one in 17 trillion. What are the real odds of someone's winning the lottery twice?**

One in 30. With millions of people buying lottery tickets, some person somewhere is bound to win it twice. Or, as a mathematician put it, "With a large enough sample, any outrageous thing is apt to happen."

The odds are still one in 17 trillion, though, against any one particular person's buying a ticket in two different lotteries and winning each time.

So, when you buy a ticket in two lotteries, the odds favor someone's eventually winning twice, but they are unbelievably stacked against *your* winning twice.

▶328 How many different ways can the cards in a deck be arranged?

Approximately 8×10^{67}, which is 8 followed by 67 zeros. Or, for those who prefer precision, the cards can be arranged 8065817 5170943878571660636856403766975289505440883277824000000 000000 different ways. Any of 52 cards could be the first in the deck; any of 51 could be second, any of 50 could be third and so on. To work it out, the number of possible arrangements is 52 factorial, which is 52 times 51 times 50, ad nauseum.

▶329 How many shuffles will mix a deck of cards thoroughly?

Seven ordinary, homestyle shuffles will thoroughly mix a pack of 52 cards. Most people, however, shuffle cards only three or four times, leaving many of the cards in their original sequence. Magicians and card cheats use this fact to improve their odds of winning.

Once the order of cards in a deck is known, shuffle and cut it three times. Then take a card out, look at it, and put it somewhere else in the deck. A magician or gambler can identify the card with near certainty!

It takes 9 shuffles to mix two decks thoroughly, and 12 shuffles to mix six decks.

▶330 When tournament bridge players switched to computerized dealing, they began getting hands with bizarrely uneven card distributions. The players suspected computer trouble. Were they right?

No, the computer was dealing the cards randomly. And a random sequence of 100 items, for example, generally includes several strings of five or six identical items.

The problem lay with the way the players had previously shuffled their cards. Bridge players group the cards in tricks of four cards, often from the same suit. When the players gathered the tricks together, shuffled them inadequately, and redealt them, the suits often wound up evenly distributed among the players. And after a while, the 4-3-3-3 or 4-4-3-2 distributions looked normal and a truly randomly generated hand looked weird.

▶331 How many perfect shuffles will thoroughly mix 52 cards? In a perfect shuffle, the cards from one hand alternate with the cards from the other hand all the way through the finished deck.

None. The cards will never be thoroughly mixed, that is, no card will be as likely to be in one place as any other card. After eight perfect shuffles, the cards are in the same order as they started before shuffling.

▶332 An apprehensive tour guide reserved hotel rooms for her travelers. Because she was in charge of an infinite busload of people, she chose the Infinity Hotel. It has an infinite number of rooms so it could guarantee every tourist a room. So why was the tour guide worried? Because she wondered, "If the hotel has an infinite number of rooms, why do I need to make any reservations at all?"

And sure enough, there was a hitch. When the bus tour arrived at the hotel and the guide asked for an infinite number of rooms, the desk clerk replied, "Sorry. They're all full."

"But you have an infinite number of rooms," the guide protested. Resourcefully, she proposed a solution. What was it?

"Move each hotel guest into a room with a number that is double the number of the room he or she started in. Then move my tourists into the emptied rooms." The hotel clerk agreed, and the guide's tired tourists tumbled happily into bed in odd-numbered rooms.

This is the basic reasoning behind the mathematics of infinities. Two infinite sets have the same cardinality—that is, they are equally numerous—if they can be paired off like the hotel rooms and guests at the Infinity Hotel.

▶333 A traveling salesman visits 100 cities a year. His penny-pinching employer wants him to travel via the most economical route, however. To oblige, the sales rep asks a computer expert to prepare the shortest possible itinerary. How long will the rep have to wait for his route?

The salesman should give up waiting and start traveling immediately. So-called traveling salesmen problems are extremely difficult to solve precisely. Even if a computer had started at the beginning of time and made a million billion calculations per second, it would have calculated the routes for only 33 cities by the year 1990. For 100 cities, the number of possible routes is about 10 to the 156th power, that is, 1 with 155 zeroes after it! No computer could operate that fast. Faced with the impossibility of obtaining precise routings, schedulers settle for approximations and manage quite nicely.

▶334 Who cares about a traveling salesman's route anyway?

Long-distance telephone companies, mail delivery systems, airlines, large factories, and computer chip manufacturers, to name a few. The so-called "traveling salesman problem" is the classic "hard" mathematics problem; the enormous number of steps involved make it mind-bogglingly complex. It is used, however, to determine the drilling order when lasers drill a million holes in a complicated computer circuit board. NASA used the traveling salesman problem to schedule astronauts' tasks in space. Factories employ it to schedule work orders in assembly lines.

▶335 How much information is contained in this Mandelbrot set?

Not much. Mandelbrot sets, after all, consist of simple calculations repeated millions of times to produce similar patterns at wildly different scales. So, despite a spectacular wealth of detail and complexity, Mandelbrot sets like this one actually rank quite low as purveyors of information—somewhere down around TV commercials and political ads.

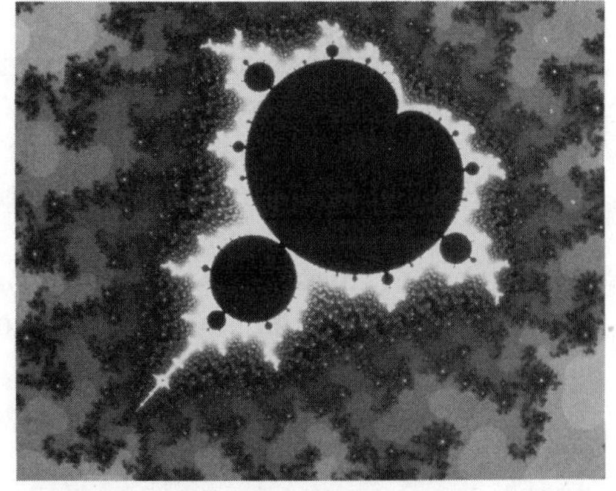

A Mandelbrot set. A Mandelbrot set contains an infinite number of similar pictures. To produce more pictures, parts of the original image can be magnified many times.

Information theory defines "information" as the length of the shortest computer program needed to produce a particular piece of data. And the computer programs that produce Mandelbrot sets are extremely short.

▶336 A lawyer with a paper shortage inadvertently caused one of the greatest mathematical mysteries of the past 350-some years. Who was the lawyer, and what was his problem?

Pierre Fermat, a French lawyer by trade, enjoyed mathematics as a hobby. In his spare time, Fermat founded number theory and was the cofounder of differential calculus, the theory of probability, and analytic geometry. His insight into the intrinsic properties of numbers is unsurpassed. Not bad for an amateur. Mathematicians call Fermat "The Prince of Amateurs."

While reading a mathematics book in 1637 Fermat had a brainstorm. The Greeks had known that many pairs of positive integers, when squared and added together, equal a third squared number. For example, $3^2 + 4^2 = 5^2$ while $5^2 + 12^2 = 13^2$. Fermat decided that he could prove that no such sum existed for cubes and higher powers. But he had run out of paper. Scribbling in the margin of the book, he wrote in Latin that, "I have discovered a truly remarkable proof, but this margin is too small to contain it."

"Fermat's Last Theorem" was neither his last nor a theorem. It was merely an unproven hypothesis. Given Fermat's track record, however, mathematicians took it seriously.

Mathematicians proved fairly easily that he was right for cubes and fourth powers all the way up to about 150,000. But there were difficulties proving or disproving Fermat's Last Theorem to infinity. In June 1993, a Princeton University mathematician, Andrew Wiles, quietly announced at the end of a lecture that he had solved the problem. Wiles had been thinking about Fermat since the age of ten. If Wiles is right, the Prince of Amateurs' reign as the great puzzler is finally over.

▶337 What's the largest known prime number?

$2^{756839} - 1$. It has 227,832 decimal digits and fills more than 46 single-spaced typed pages. It was produced by multiplying 2 by itself 756,839 times and then subtracting 1. There is no "largest" prime number, because infinitely many primes exist. Thus, this is merely the largest *known* prime number.

Prime numbers are the atoms of mathematics, the building blocks of the number system. They are not evenly divisible by any positive integer except themselves and one. Puny examples include 2,3,5,7, and 11.

Mathematicians have had to invent new shortcuts in multiplication to test enormous numbers for primality. Without new techniques, a computer capable of doing a billion divisions a second would have needed more than 10^{60} years to factor a 155-digit number. Weather forecasters, oil prospectors, researchers of fluid flow aerodynamics, astronomers searching for pulsars, and others made immediate use of the improved multiplication methods.

▶338 The news that mathematicians had broken a 100-digit number into two prime numbers shocked the world of cryptography in 1988. Why the uproar?

Cryptographers for large institutions such as banks, telephone companies, and the military use enormous products of prime numbers to encode their data. Prime numbers are numbers that cannot be divided any further without leaving a remainder. Until 1988, anyone could use a 100-digit number to encrypt data. Without the key to its prime factors, no one could decode and retrieve the information. Hence, a 100-digit number was considered safe. As late as 1971, mathematicians could factor only 40-digit numbers.

Teams of mathematicians used 400 computers in the United States, the Netherlands, and Australia to crack the 100-digit number. Together, they factored it into two prime numbers 41 and 60 digits long. Multiplying those two numbers together produces the 100-digit number.

As difficult as factoring huge numbers is, it is still easier than solving many-route traveling salesmen problems. And by 1990, several hundred mathematicians and a thousand computers had factored a difficult 150-digit number into three primes. That number, large though it was, was actually easier to factor than some much smaller numbers.

▶339 What is the largest cryptographer's key that has been factored?

A 120-digit number that was factored into two prime numbers, each with 60 digits. Many larger numbers have been broken down into their factors, but those numbers were not as mathematically challenging. When a group of cryptographers issued a challenge to the mathematics community to try to factor a list of difficult numbers used as cryptography keys, the 120-digit number was the third smallest number on the list. The number was factored on June 9, 1993. The number is

227010481295437363334259960947493668895875336466084780038173258247009162675779735389791151574049166747880487470296548479

which equals

327414555693498015751146303749141488063642403240171463406883

times

693342667110830181197325401899700641361965863127336680673013.

▶340 The general public knows pi—the ratio of a circle's circumference to its diameter—as 3.14. For most scientific applications, ten decimal digits are enough. What is the record for the number of decimal places for pi?

Computer technology has spawned the mathematics of huge numbers. The record-holders for pi are David V. and Gregory U. Chudnovsky, two brothers at Columbia University. They produced 2.260+ billion decimal digits in September 1991. If the numbers were lined up, they would stretch all the way across the continental United States and into the Pacific.

Pi has become the ultimate test of the speed, accuracy, and efficiency of new computers and their programs. An error in even one of pi's million digits points to problems in either the computer or its programming. Pi is also considered an excellent generator of random numbers.

▶341 "Publish or perish" normally means publication in a scholarly journal. What mathematical discovery was first published in a novel?

A new way to construct numbers, discovered by John Horton Conway and explained briefly to Donald Knuth, who wrote a novel about it. *How Two Ex-Students Turned on to Pure Mathematics and Found Total Happiness* was published by Addison-Wesley in 1974.

Incidentally, Knuth is better known as a computer science scholar than as a novelist.

▶342 "How I want a drink, alcoholic of course, after the heavy lectures involving quantum mechanics!" the math student muttered. What precisely does the distraught student really want?

A. Beer.
B. Wine.
C. Whiskey.
D. Pi (e).

D, Pi. Count the number of letters in each word of the sentence and arrange the numbers in order. This mnemonic device won't get a math student a drink, but it does produce the approximate value of pi: 3.14159265358979.

▶343 Do hot shots have hot hands? According to basketball lore, a player who has made one successful shot is more likely to be successful on the second throw because he has "hot hands." Are hot hands full of hot air?

Yes, according to a Stanford University psychologist who analyzed every shot made by the Philadelphia 76ers for one and a half seasons. The sad reality was that a 76er who made one successful shot was a bit more likely to *miss* on the second try.

▶344 Flip a coin 100 times and watch the same side come up five or six times in a row. Is the coin weighted? Or is it random?

It is truly random. In any 100 coin toss, several streaks of five or six identical sides are likely to occur. When trying to write a "random" list of numbers, people almost invariably mix the numbers too much.

Producing a truly random sequence of numbers is extremely difficult, even for computers. Modern society—from state lotteries to researchers, juries, and the like—relies on getting random numbers to ensure fairness. Yet not even the most advanced computer program provides completely random sequences. Computer programs, by their very nature, are predictable. Thus, it is almost impossible for them to make a completely unpredictable product.

▶345 You're a contestant on the "Let's Make a Deal" TV game show. Host Monty Hall shows you three doors: one has a Corvette behind it and the other two hide goats. Hall knows which door has the car, but you don't. But if you pick it, it's yours.

After you choose the first door (Door 1), Monty Hall opens one of the other doors to show you a goat. (We assume that Monty Hall is impartial and is making random choices.) Eliminating Door 2 with the goat narrows your final choice to two doors: your original selection (Door 1), or the remaining door, Door 3.

Should you switch your choice from Door 1 to Door 3?

Yes. When Marilyn vos Savant posed the problem in her *Parade* magazine column, she used her 228-point IQ to advise contestants to switch. Many mathematicians disagreed, and vos Savant received more than 10,000 irate letters. The controversy raged for weeks. It was hotly debated in the Central Intelligence Agency, the Massachusetts Institute of Technology, the Los Alamos National Laboratory in New Mexico, and in hundreds of classrooms.

As Monty Hall later explained, the first time the contestant chooses a door, the odds of winning the car are one in three. Opening Door 2 does not change the odds on Door 1; those remain one in three. But eliminating Door 2 as a possibility does change the odds of finding a car behind the remaining Door 3. Your chance of discovering a car behind Door 3 becomes two out of three. So it pays to switch. And Marilyn vos Savant was right.

▶346 An all-too-typical dinner party crisis involves seating couples at the table so that men and women alternate and couples don't sit next to each other. Known mathematically as the Ménage Problem, it took mathematicians a half century to solve it and another half century to understand their solution. The obstacle, it seems, was sexism. Explain, please.

Traditionally, the problem was solved by politely seating "ladies first." Only in 1986 did mathematicians realize that a gender-free approach produces a simpler and more elegant two-step solution. First, break the couples apart and seat them regardless of their sex and only later alternate the men and women.

Solving and explaining the Ménage Problem ended the mathematical confusion but merely confirmed the plight of the hapless host and hostess: With three couples, there are only 12 suitable seating arrangements. But with four couples, there are 96. And with five couples, there are 3,120 bewildering possibilities, and with ten couples a mind-boggling 3,191,834,419,200 choices!

▸347 A banquet organizer must arrange the seating for 150 guests. But some of the guests hate each other while others are close friends. How should they be seated?

Resign yourself to some angry guests, because the problem is unsolvable. "Seating arrangement" problems are even more difficult than traveling salesmen problems. If you wrote each possible seating arrangement on a card and lined all the cards up, they would reach beyond the universe.

▸348 A friend invites you to a party and then tries to bet you that two of the guests will share the same birth date. Should you take the bet?

Obviously, if 366 people attend the party, he has a sure bet and he's not a good friend. To make the bet fair, how small can the party be (assuming, of course, that he hasn't been studying birth records on the sly)?

23. According to probability, the chances are 50-50 that two people in any group of 23 will share the same birthday. With only 14 people in the group, there is a 50-50 chance of finding two persons born within a day of each other. And you need only seven in the group to find a 50-50 chance that two people have birthdays within a week of each other.

The point is that most coincidences turn out to be quite predictable.

▸349 What is the most important mathematical result ever patented?

A mathematical discovery that greatly speeds the solution of extremely difficult scheduling problems, like the routing of telephone calls. The American Telephone and Telegraph Company patented its algorithm, the set of computer directions for carrying out the solution, in 1988. Many mathematicians were incensed, because they believe that patenting an abstract procedure impedes the free flow of ideas.

AT&T's next challenge will be legal, not mathematical: defending the patent in court.

▶350 In 1930, classicists celebrated the 2,000th anniversary of the birth of Virgil, Rome's greatest poet and the author of the epic *Aeneid.* Partway through the festivities, a mathematician was graceless enough to point out that the 2,000th anniversary of Virgil's birth in 70 B.C. would not occur until a year later in 1931. How had the scholars miscalculated?

When counting years, they had forgotten that Virgil was born before Christ. The years B.C. pass to A.D. without going through zero. The year before A.D. 1 is 1 B.C. Adding an extra year to compensate for the lost zero shifted the august Augustan's birthday a year later. A related problem will occur in the year 2000 when many people will insist that the year 2000 is the first year of the twenty-first century instead of the last year of the twentieth century: a confusion over naught! Or, as Shakespeare put it, *Much Ado about Nothing.*

▶351 What opiate-addicted titled lady, the daughter of a dashing poet, first described the computer?

Augusta Ada Byron, the Lady Lovelace. She helped finance and program Charles Babbage's Analytical Engine. In 1843, she also became the first to describe the machine, considered the world's first computer.

Lady Lovelace was the mathematically inclined friend of Babbage, Charles Dickens, and Michael Faraday. In the early 1980s, the U.S. Department of Defense named its computer programming language "Ada," in her honor.

▶352 The following diagram represents a new mathematical principle. What is it?

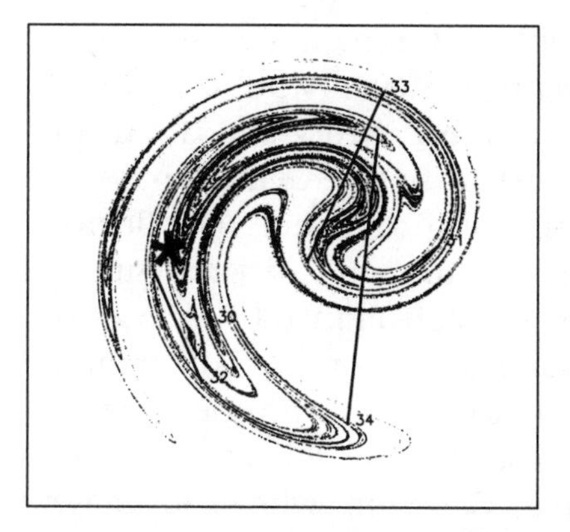

Chaos. Chaotic systems like the weather or turbulent water are inherently unpredictable because they involve stretching and folding. Stretching quickly magnifies the slightest difference between the initial values, and folding brings the system back within bounds. The starting values (at the *) in this "Ikeda map" diagram differ only in the eighth decimal place. Substituting the initial values into the Ikeda equation over and over again each time it is solved produces increasingly divergent values. The ends of the numbered lines show how far apart the original values have grown by the 31st, 32nd, 33rd, and 34th calculations.

▶353 **George Dantzig, who invented the simplex method of linear programming, almost flunked junior high mathematics. In graduate school, he arrived late for a class, copied down two problems from the blackboard as his homework assignment, and struggled for several days before completing them. Then, apologizing for handing his work in late, he left his solutions on the professor's desk.**

One Sunday morning several weeks later as Dantzig was eating breakfast, the professor banged on his front door and dashed inside. What had Dantzig done to agitate the professor so?

The two problems on the blackboard had not been homework. They were two famous, unsolved problems in statistics. And Dantzig had solved them. The professor barged in on Dantzig's breakfast, shouting, "I've just written an introduction to one of your papers. Read it so I can send it out right away for publication."

▶354 **Rubbing a bow against a plate covered with sand makes mathematical "sand pictures." The sand bounces away from vibrating regions and collects at "nodes," which remain still. Depending on the shape of the plate, position of its supports, and the frequency of the vibration, different patterns are created. Who explained the physics and mathematics of the patterns?**

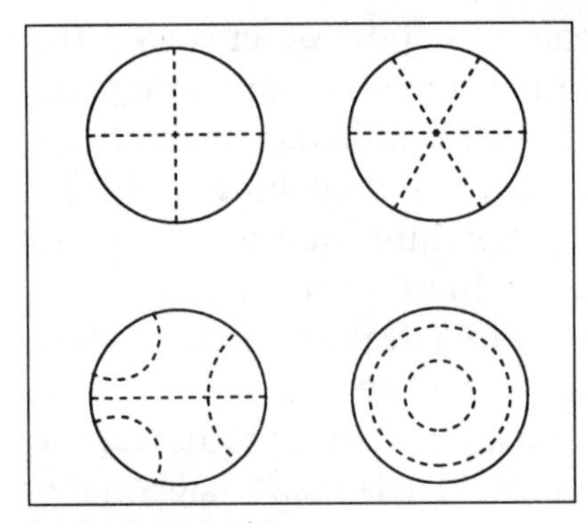
Sand patterns on plates.

Sophie Germain, a self-taught mathematician who contributed to number theory and the theory of elasticity. The German physicist Ernst F. F. Chladni had demonstrated his sand experiments before Napoleon in 1808. Impressed, Napoleon offered a kilogram of gold to anyone who could explain mathematically what was happening. Germain, who had taught herself mathematics and physics from lecture notes and books, won the prize in 1815. During her lifetime she exchanged letters on number theory with Carl Friedrich Gauss, Joseph-Louis Lagrange, and Jean-Baptiste-Joseph Fourier. Gauss tried, without success, to get her a degree from the German University of Göttingen.

▶355 What was the world's first video proof of a mathematical theorem?

"Not Knot," a 15-minute video created on a computer to prove with pictures a special case of a topological problem called the Thurston conjecture. The video was created by Charlie Gunn at the Geometry Center in Minneapolis.

Computers have opened up mathematics to new approaches like this video experiment with pictures. Computers have even created "experimental mathematicians." Traditionally, mathematicians have used logical deduction to prove general principles about phenomena. "Experimental mathematicians," on the other hand, experiment with computers to find patterns in abstract worlds created by their computers. They search for practical evidence that a phenomenon actually exists in nature.

▶356 What is the most economical way to pack oranges into a box?

To pack the most spheres into the smallest possible space, greengrocers pile oranges in layers so that the fruits nestle into gaps between the four spheres in the layer below. But mathematicians wondered if

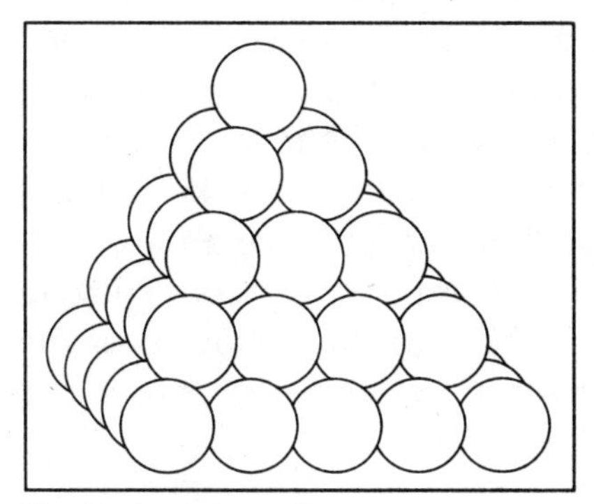

The classic orange stack.

this also held true for an infinite number of spheres in infinite space. It has been proven that in a thousand-dimensional space (rather than a three-dimensional space), the greengrocer's orange-stack is not the densest pack. Supermarkets will be surprised to know that this issue is one of the oldest puzzles in mathematics. It was first raised by Johannes Kepler in the seventeenth century.

▶357 In the arcane world of American measurements, which weighs more:

- **An ounce of gold or an ounce of lead?**
- **A pound of gold or a pound of lead?**
- **A pound of feathers or a pound of gold?**

Thanks to the failure of the U.S. government to adopt the metric system, an ounce of gold is heavier than an ounce of lead, although a *pound* of gold is lighter than a pound of lead. And just to make you mad enough to write your Congressman (or woman), a pound of feathers is heavier than a pound of gold.

Gold is measured in troy pounds whereas lead and feathers are measured in avoirdupois pounds. Troy weights originated in the French city of Troyes for gold, silver, and precious stones. Troy pounds have 12 ounces; avoirdupois pounds have 16 ounces. But a troy pound (at 372 grams in the metric system) weighs less than an avoirdupois pound (which is 454 metric grams). Thus, a troy pound is eight-tenths of an avoirdupois pound. On the other hand, each troy ounce is heavier than an avoirdupois ounce. Ready for metric?

▶358 Fuzzy logic is:

A. Woolly-headed thinking.
B. A warm and user-friendly computer program.
C. A computing system that mimics human reason.

C, a computer programming system that helps computers make decisions like people. When input data are vague and mathematical models are nonexistent or too complex, people in the real world still manage to come up with answers. Their information may be ambiguous, contradictory, or imprecise, but they can produce an approximate answer. Now, thanks to fuzzy logic, computers can too. They no longer require information to be one thing to the exclusion of the other; now data can be partly one thing and partly another too.

▶359 Traceries, the decorations carved around windows in Gothic cathedrals, are one of the few Western ornaments that owe nothing to the Greeks and Romans. Why would the stonemasons who created them consider themselves scientists, several social steps above mere artisans?

Traceries are based on geometry, particularly circles with triangles inside them. A circle is divided into five equilateral triangles with curved sides in a window in the left photo below from the Cistercian Cloisters of Hauterive south of Fribourg, Switzerland. Then each triangle is further subdivided into three more, and so on. Each of the 20 cloister windows at Hauterive represents a different geometric pattern. The windows illustrate Book IV of Euclid's *Elements*. In it, Euclid shows how to construct within circles regular polygons with 3, 4, 5, 6, and 15 sides.

Most traceries are based on much simpler geometry, however. The circle is divided into six equilateral triangles in the tracery at Zistertiesuser Church in Zwett, Austria (see right photo).

Window traceries at the Cistercian Cloisters of Hauterive, Switzerland (left), and the Cistercian Church in Zwett, Austria (right).

A Final, Interdisciplinary Quiz

To emphasize the interdisciplinary character of modern science, here is a quick quiz across scientific lines. Measure the breadth of your scientific knowledge here.

▶360 **What scientific theories did these famous cruises prove?**

A. ***Beagle.***
B. ***Glomar Challenger.***
C. An eclipse-viewing cruise to the Gulf of Guinea in 1919.
D. Edmond Halley's 1676 trip to Saint Helena.

A. Evolution. Charles Darwin's five-year trip on the HMS *Beagle* to survey South American and Pacific island wildlife convinced him that species of life on Earth had evolved gradually. He published his theory in *On the Origin of Species* more than 20 years later in 1859.

B. Continental drift. Core samples gathered between 1968 and 1983 by the *Glomar Challenger,* an oceanographic drilling and coring ship, established that the seafloor is spreading and that the continents are drifting apart.

C. General theory of relativity. Calculations of the solar eclipse made in the Gulf of Guinea verified Einstein's general theory of relativity. Observations of starlight passing by the sun during the total eclipse confirmed his prediction that light from distant stars is deflected in the gravitational field of the sun.

D. Gravity. The English astronomer Edmond Halley's pendulum measurements at Saint Helena in the southern Atlantic helped Isaac Newton formulate the theory of gravitation in 1687. Halley's data must have been good, because Newton's theory is still valid for all but the most precise measurements. Later, Halley published Newton's masterwork, *Principia,* about celestial mechanics.

▶361 Magnolia and gingko trees, tapirs, Australian lungfish, and the nautilus are virtually the same today as their fossilized ancestors were eons ago. They are called "living fossils." The aristocrats of living fossils, however, were known to science first as fossils and only later discovered alive and well in the wild. What are the two granddukes of the fossil world?

Dawn redwood trees and the coelacanth fish. A feathery conifer, the dawn redwood (*Metasequoia glyptostroboides*) was discovered in 1941 in fossils from Japan. Dawn redwoods dominated northern forests for millions of years before the last ice age. Until 1946, the species was considered extinct. Then it was found growing in central China near the Yangtze River (Jinsha Jiang). That was its last natural refuge.

The coelacanth *Latimeria chalumnae* is a common fossil in sedimentary rocks dating from 400 million to 65 million years ago. In 1938, fishermen caught a living coelacanth in the Indian Ocean off South Africa. In the late 1980s, scientists in a submersible craft filmed coelacanths swimming in their natural habitat about 200 meters underwater.

▶362 The boundaries between traditional scientific categories are blurring. What inhabits the areas between the following categories?

A. Liquids and solids.
B. Atoms and bulk matter.
C. Organic and inorganic chemistry.

A. Liquid crystals. They are liquids, but their molecules are arrayed in an orderly arrangement like a crystal. Discovered in 1888 by an Austrian botanist, they remained curiosities until the 1960s. Since then, they have been used for wristwatch displays, ultra-high-strength fibers like Kevlar, and computer screens.

B. Clusters. Clusters are groups of hundreds or thousands of atoms. Sometimes called the "fifth state of matter," they have become a hot topic of study by physicists and chemists.

C. Organometallic chemistry, which studies compounds like vitamin B-12 in which carbon and metal atoms are directly bonded together. Traditionally, inorganic chemists studied inanimate metals and minerals while organic chemists studied carbon-containing compounds from living organisms.

►363 **America, the land of the immigrants, has had some unwelcome invaders. Foreign plants and animals may arrive without the predators that kept them in check back home. Then the newcomers can overwhelm native species that are not adapted to the foreign organisms.**

Large freighters use ballast water and sand for balance and stabilization. When they take on cargo, they release ballast water from one ocean into another. The practice launched one of the most successful invasions ever of the United States. What was the invader and the territory it occupied?

A zebra mussel cluster.

The zebra mussel, the most notorious immigrant into North American freshwaters. Since 1988, the zebra mussel has muscled its way aboard large freighters from Europe into most of the waterways in the United States via the St. Lawrence Seaway and the Great Lakes. With no local predators, it has squeezed out many native species. It is only one of 136 exotic plants and animals to have settled in the Great Lakes, however. More than a third of them have arrived since the St. Lawrence Seaway opened in 1959.

►364 **What plant, introduced to the northwestern United States as oyster packing, is reshaping coastlines there?**

Spartina, a tall and graceful cordgrass common to the salt marshes of the East and Gulf Coasts. In 1894, it cushioned crates of oysters being transplanted from the East Coast to the state of Washington. Now it has spread, threatening Washington's oyster harvests, which are the second-largest in the nation. The plant, which can trap 6 inches of sediment yearly, transforms mudflats into high marshes. Mudflats are much more hospitable to many species of fish, shellfish, and birds.

▶365 Three famous scientists—Newton, Gregor Mendel, and Robert Millikan—stand accused today of violating modern scientific standards. What did they allegedly do?

Fudge their data.

Newton, one of the fathers of modern science, massaged his mathematical equations and experimental data to bolster his theories. Technically, no one in the seventeenth century could have worked out the law of universal gravitation to a precision of one part in 30,000, as he claimed to have done. Newton also ordered his editor to "mend the numbers" in the *Principia* so that the Earth's precession of equinoxes would come out to the proper answer of 50 seconds.

Gregor Mendel's results are also too good to be true. Mendel, the father of modern genetics, classified garden peas by their inherited characteristics. When some of the peas did not fit clearly into any of his categories, either Mendel or his assistant put the doubtful peas into the groups that would best support his theories.

Robert Millikan won a Nobel Prize in 1910 for his oil drop experiment, which determined the electrical charge of the electron. As the second American physicist to win a Nobel, Millikan was a powerful figure in U.S. science for decades afterwards. But Millikan had reported only two-thirds of his data—the parts that fit his theory. He claimed that he had reported "*all* the results experimented on." An Austrian physicist who performed the same experiment did report all *his* data and consequently got much less spectacular results. Thanks to his frankness, he missed out on the Nobel Prize. Failing to report data would be considered unethical misconduct today.

Recommended Reading

The numbers correspond to the numbers of the "facts" in the book.

Engineering and Technology

1 Jonathan Coopersmith. "Facsimile's False Starts." *IEEE Spectrum* 30 (Feb. 1993): 46–49; Tim Hunkin. "Just Give Me the Fax." *New Scientist* 137 (Feb. 13, 1993): 33–37.

2 Sebastian Junger. "The Pumps of New Orleans." *Invention and Technology* 8 (Fall 1992): 42–48.

3 Renso Gasparotto. "Waterfall Aeration Works." *Civil Engineering*, Oct. 1992, 52–54.

4 David Bjerklie. "High-Tech Olympians." *Technology Review*, Jan. 1993, 25–30.

5 Henry Petroski. *To Engineer is Human: The Role of Failure in Successful Design*. New York: Random House, 1982.

6 American Chemical Society. *Science and Serendipity: The Importance of Basic Research*. Washington, D.C.: American Chemical Society, 1992.

7 Christopher Viney, Anne E. Huber, and Pedro Verdugo. "Liquid Crystalline Order in Mucus." *Macromolecules* 26 (Feb. 1993): 852–855.

8 Christopher Viney, lecture, University of Washington; Mehmet Sarikaya and Ilhan A. Aksay. "Nacre of Abalone Shell: a Natural Multifunctional Nanolaminated Ceramic-Polymer Composite Material." In *Structure, Cellular Synthesis and Assembly of Biopolymers*, ed. St. T. Case. Berlin: Springer-Verlag, 1992.

9 Charles Singer et al. *A History of Technology*. Vols. II and IV. Oxford: Clarendon Press, 1956 and 1958.

10 K. Kerkam, C. Viney, D. Kaplan, and S. Lombardy. "Liquid Crystallinity of Natural Silk Secretions." *Nature* 349 (Feb. 14, 1991): 596–598; Fritz Vollrath. "Spider Webs and Silks." *Scientific American* 266 (March 1992): 70–76.

11 Robert Mark. "Lessons from the Master Builders." *1991 Yearbook of Science and the Future, 1991 Yearbook*. Chicago: Encyclopaedia Britannica, Inc., 1991.

12 Chester R. Kyle. "Athletic Clothing." *Scientific American* 254 (March 1986): 104–110.

13 Michael Gianturco. "The Infinite Straightaway." *Invention and Technology* 8 (Fall 1992): 34–41.

14 John M. Coles, "The World's Oldest Road." *Scientific American* 261 (Nov. 1989): 100–106.

15 Anne Simon Moffat. "Engineering at the Lower Limits of Size." *Mosaic* 21 (Winter 1990): 30–40.

16 John Reganold. "Farming's Organic Future." *New Scientist* 122 (June 10, 1989): 49–52.

17 Mick Hamer. "Trains That Go Pop in the Dark." *New Scientist* 123 (Sept. 9, 1989): 63–65.

18 David Anthony, Dimitri Y. Telegin, and Dorcas Brown. "The Origin of Horseback Riding." *Scientific American* 266 (Dec. 1991): 94–100.

19 See 18.

20 Randall R. Inouye and Joseph D. Jacobazzi. "The Great Chicago Flood of 1992." *Civil Engineering*, Nov. 1992, 52–55.

21 Charles R. Heidengren. "Settling Down Easy." *Civil Engineering*, Dec. 1992, 72–74.

22 Michael Valenti. "Tapping Landfills for Energy." *Mechanical Engineering,* Jan. 1992, 44–47.

23 R. A. Tikhomirov et al. "High-Pressure Jetcutting." *Mechanical Engineering,* June 1992, 88–91; Christian M. Olsen and Robert H. Todd. "Designing and Building a Water Jetcutting Machine." *Mechanical Engineering,* July 1992, 68–72.

24 Vilma Barr. "Alexandre Gustave Eiffel: A Towering Engineering Genius." *Mechanical Engineering,* Feb. 1992, 58–65.

25 Fred Hapgood. "The Really Little Engines That Might." *Technology Review* 93 (Feb./March 1993): 31–36; Henry I. Smith and Dimitri A. Antoniadis. "Seeking a Radically New Electronics." *Technology Review* 93 (April 1990): 26–40.

26 See 15; and D. M. Eigler and E. K. Schweizer. "Positioning Atoms with a Scanning Tunnelling Microscope." *Nature* 344 (April 5, 1990): 524–526.

27 See 25.

28 Tabitha M. Powledge. "Gene Pharming." *Technology Review* 95 (Aug./Sept. 1992): 61–66.

29 A. A. Wessol and Bill Whitacre. "Operating Hydraulics on 'Green' Fluids." *Machine Design,* Jan. 22, 1993, 73–77.

30 Joe P. Mahoney, University of Washington Civil Engineering Department, private communication.

32 Joe P. Mahoney, University of Washington. Draft, *Washington State Department of Transportation Pavement Guide* Vol. I, Jan. 1993; Igal Talmi, private communication.

33 See 32.

34 Victoria and Dale Lightfoot. "Revealing the Ancient World Through High Technology." *Technology Review* 92 (May/June 1989): 54–61.

35 Tekla S. Perry. "Cleaning Up." *IEEE Spectrum* 30 (Feb. 1993): 20–26.

36 See 35.

37 Fred Pearce. "Ancient Lessons from Arid Lands." *New Scientist* 132 (Dec. 7, 1991): 42–48.

38 See 37.

39 Malcolm W. Browne. "Doomed Highway May Have Pulsed to the Rhythm of the Deadly Quake." *New York Times*, Oct. 23, 1989, B10.

Medicine, Health, and Nutrition

40 Roy M. Anderson and Robert M. May. "Understanding the AIDS Pandemic." *Scientific American*, May 1992, 58–66; Anne Simon Moffatt. "Models and Theory for an Ecological Phenomenon." *Mosaic* 20, no. 1 (Spring 1989): 4–15.

41 Joseph Alper. "Ulcers as an Infectious Disease." *Science* 260 (April 19, 1993): 159–160.

42 David Krogh. *Smoking, the Artificial Passion*. New York: W. H. Freeman and Co., 1991.

43 Barry R. Bloom and Christopher J. L. Murray. "Tuberculosis: Commentary on a Reemergent Killer." *Science* 257 (Aug. 21, 1992): 1055–63.

44 Paul Harvey and Robert May. "Matrimony, Mattresses and Mites." *New Scientist*, March 3, 1990, 48ff.

45 Robin Marantz Henig. "Asthma Kills." *New York Times Magazine*, March 28, 1993, 42ff.; and see 44.

46 Michael Stroh. "The Root of Impotence: Does Nitric Oxide Hold the Key?" *Science News* 142 (July 4, 1992): 10–11; Louis J. Ignarro. "Nitric Oxide as the Physiological Mediator of Penile Erection." *Journal of NIH Research* 4 (May 1992): 59–62; and Arthur L. Burnett et al. "Nitric Oxide: A Physiologic Mediator of Penile Erection." *Science* 257 (July 17, 1992): 401–494.

47 Associated Press. "Why a Beer Belly Is Precisely That." *New York Times*, April 9, 1992, A14; Paolo M. Suter, Yves Schutz, and Eric Jequier. "The Effect of Ethanol on Fat Storage in Healthy Subjects." *New England Journal of Medicine* 326, no. 15 (April 9, 1992): 983–987.

48 See 43.

49 "Feeling Down? Try Eating." *Technology Review* 93 (May/June 1990): 14.

50 Rose E. Frisch. "Fatness and Fertility." *Scientific American* 258 (March 1988): 88; Rose E. Frisch, ed., *Adipose Tissue and Reproduction*. Basel, Switzerland: S. Karger Publishers, 1990; and Rose E. Frisch. "Body Fat, Menarche and Fertility." *Encyclopedia of Human Biology* I, 741–750. New York: Academic Press, 1991.

51 Robin Marantz Henig. *A Dancing Matrix: Voyages Along the Viral Frontier*. New York: Alfred A. Knopf Inc., 1992; Robin Marantz Henig. "Flu Pandemic." *New York Times Magazine*, Nov. 29, 1992, 28ff.

52 Colin McEvedy. "The Bubonic Plague." *Scientific American* 258 (Feb. 1988): 118–123; Charles L. Mee, Jr. "How a Mysterious Disease Laid Low Europe's Masses." *Smithsonian* 20 (Feb. 1990): 66–74.

53 "Babies Fuss over Postexercise Milk." *Science News*, July 18, 1992, 47.

55 *Health United States 1991*. U.S. Department of Health and Human Services, p. 161.

56 "Boom and Doom." *Harvard Health Letter* 16 (Dec. 1990): 1–4.

57 Philip A. Mackowiak, Steven S. Wasserman, and Myron M. Levine. "A Critical Appraisal of 98.6 Degrees F, the Upper Limit of the Normal Body Temperature, and Other Legacies of Carl Reinhold August Wunderlich." *JAMA* 268, no. 12 (Sept. 23/30, 1992): 1578–1580.

58 Timothy Johns. "Well-Grounded Diet." *The Sciences* 31 (Sept./Oct. 1991): 39–43; Timothy Johns. *With Bitter Herbs They Shall Eat It: Chemical Ecology and the Origins of Human Diet and Medicine*. Tucson: University of Arizona Press, 1990.

59 Charles E. Rosenberg. *The Care of Strangers: The Rise of America's Hospital System*. New York: Basic Books Inc., 1987.

60 Natalie Angier. "A Potent Peptide Prompts an Urge to Cuddle." *New York Times*, Jan. 22, 1991, C1; C. Sue Carter and Lowell L. Getz. "Monogamy and the Prairie Vole." *Scientific American* 268 (June 1993): 100–106.

61 James E. Dalen. "An Apple a Day or an Aspirin a Day?" *The Archives of Internal Medicine* 151 (June 1991): 1066–1068.

62 *Abstract of the United States 1992*. U.S. Department of Commerce, Bureau of the Census; and the Sierra Regional Eye and Tissue Bank, University of California Davis Medical Center.

63 "If Queen Victoria Had Known about LAM." *The Lancet* 337 (March 23, 1991): 703.

64 Ronald Finn. "John Bostock, Hay Fever, and the Mechanism of Allergy." *The Lancet* 340 (Dec. 12, 1992): 1453–1454.

65 Nicholas D. Kristof. "Stark Data on Women: 100 Million Are Missing." *New York Times*, Nov. 5, 1991, C1.

66 James C. Whorton. "Inner Hygiene: The Philosophy and Practice of Intestinal Purity in Western Civilization." In *History of Hygiene, Proceedings of the 12th International Symposium on the Comparative History of Medicine—East and West* 1987. Tokyo: Ishiyaku EuroAmerican, Inc., 1991; James C. Whorton. *Crusaders for Fitness: The History of American Health Reformers*. Princeton: Princeton University Press, 1982.

67 Samuel M. Lesko, Lynn Rosenberg, and Samuel Shapiro. "A Case-Control Study of Baldness in Relation to Myocardial Infarction in Men." *JAMA* 269 (Feb. 24, 1993): 998–1003.

68 T. P. Coultate. *Food: The Chemistry of Its Components*. London: Royal Society of Chemistry, 1989.

69 See 68.

70 Lewis L. Nielsen. "Mosquitoes Unlimited." *Natural History* 100 (July 1991): 4ff.

71 Abraham Flexner. *Medical Education in the United States and Canada*. Bulletin no. 4. New York: Carnegie Foundation for the Advancement of Teaching, 1910; Paul Starr. *The Social Transformation of American Medicine*. New York: Basic Books, Inc., 1982.

72 Jane E. Brody. "To Preserve Their Health and Heritage, Arizona Indians Reclaim Ancient Foods." *The New York Times*, May 21, 1991, C1.

73 Barry Bogin. "Why Must I Be a Teenager At All?" *New Scientist* 137 (March 6, 1993): 34–38.

74 See 73; Barry Bogin. *Patterns of Human Growth.* Cambridge: Cambridge University Press, 1988; and Marquisa LaVelle Moerman. "Growth of the Birth Canal in Adolescent Girls." *American Journal of Obstetrics and Gynecology* 143 (1982): 528–532.

75 Jane E. Brody. "Huge Study of Diet Indicts Fat and Meat." *New York Times,* May 8, 1990, C1.

76 F. John Ebling. "Biology of Hair Follicles." In *Dermatology in General Medicine,* Thomas B. Fitzpatrick et al. New York: McGraw-Hill, 1993.

77 Eugene N. Straus and Rosalyn S. Yalow. "Immunochemical Studies Relating to Cholecystokinin in Brain and Gut." In *Recent Progress in Hormone Research,* issued by Laurentian Hormone Conference, ed. Roy O. Greep. New York: Academic Press, 1981.

78 Fernando Nottebohm. "From Bird Song to Neurogenesis." *Scientific American* 260 (Feb. 1989): 74–79.

79 Edward R. Wolpow. "After the Fall." *Harvard Health Letter* 16 (April 1991): 1–3; B. Bower. "Clues to the Brain's Knowledge Systems." *Science News* 142 (Aug. 29, 1992): 148.

80 Vilayanur S. Ramachandran. "Perceiving Shape from Shading." *Scientific American* 259 (Aug. 1988): 76–83; Vilayanur S. Ramachandran. "Blind Spots." *Scientific American* 266 (May 1992): 86–91; and Dorothy A. Kleffner and V. S. Ramachandran. "On the Perception of Shape from Shading." Perception & Psychophysics *52,* no. 1 (1992): 18–36.

81 Elisabeth Rosenthal. "4,000-Year-Old Treatment Revived to Heal Wounds." *New York Times,* April 5, 1990, B7.

82 See 81.

83 Susan Katz Miller. "Cells into Organs." *Technology Review* 95 (July 12, 1992): 12–13.

84 C. Ezzell. "Pig Intestine Yields Versatile Tissue Graft." *Science News* 141 (April 25, 1992): 246.

85 Elisabeth Rosenthal. "Doctors Trying Coral for Skeletal Repairs." *New York Times,* July 2, 1991, C3.

Ecology and Animal Behavior

86 "A Butterfly Taste Test Discredits Old Theory." *New York Times*, April 16, 1991, C3; David B. Ritland and Lincoln P. Bower. "The Viceroy Butterfly Is Not a Batesian Mimic." *Nature* 350 (April 11, 1991); 497–498.

87 Helmut Hemmer. *Domestication: The Decline of Environmental Appreciation*. Cambridge: Cambridge University Press, 1990.

88 Laurence G. Frank and Stephen E. Glickman. "Born to Kill"; and Laura Smale and Kay E. Holekamp. "Growing Up in the Clan." *Natural History* 102 (Jan. 1993): 42–52.

89 Edward O. Wilson. *The Diversity of Life*. Cambridge: Harvard University Press, 1992.

90 Kathryn A. Minta and Steven C. Minta. "Partners in Carnivory." *Natural History* 100 (June 1991): 60–62.

91 Andrew Cockburn and Anthony K. Lee. "Marsupial Femmes Fatales." *Natural History* 97 (March 1988): 41–46.

92 John T. Hardy. "Where the Sea Meets the Sky." *Natural History* 100 (May 1991): 59–65.

93 Paul W. Sherman, Jennifer U. M. Jarvis, and Stanton H. Braude. "Naked Mole Rats." *Scientific American* 267 (August 1992): 72–78; Richard W. Sherman, Jennifer U. M. Jarvis, and Richard D. Alexander, eds. *The Biology of the Naked Mole-Rat*. Princeton: Princeton University Press, 1991.

94 "Endangered Parrot Lays Nest Egg for Survival." *New Scientist* 262 (March 10, 1990): 27; Don Merton, Kakapo Project Leader, Threatened Species Unit, Department of Conservation, Wellington, New Zealand. Private communication, June, 1993.

95 See 89.

96 Fiona King. "Mind the Bugs Don't Bite." *New Scientist* 262 (Jan. 27, 1990): 51–54.

97 Bruce M. Beehler. "The Birds of Paradise." *Scientific American* 261 (Dec. 1989): 117–123.

98 Gerald S. Wilkinson. "Food Sharing in Vampire Bats." *Scientific American* 262 (Feb. 1990): 76–82.

99 Wallace Arthur. "The Bulging Biosphere." *New Scientist* 130 (June 29, 1991): 42–45; Wallace Arthur. *The Green Machine: Ecology and the Balance of Nature*. London: Basil Blackwell, 1991; and Patrick Huyghe. "New-Species Fever." *Audubon* 95 (March-April 1993): 88–96ff.

100 See 99, Arthur; and 89.

101 See 89.

102 Bernd Heinrich and George A. Bartholomew. "The Ecology of the African Dung Beetle." *Scientific American* 241 (Nov. 1979): 146–149ff.

103 Christoph Boesch and Hedwige Boesch-Achermann. "Dim Forest, Bright Chimps." *Natural History* 100 (Sept. 1991): 50–56.

104 William K. Stevens. "To Drive Away Wasps, Caterpillars Recruit a Phalanx of Ants." *New York Times*, Aug. 6, 1991, C1; Philip J. DeVries. "Singing Caterpillars, Ants, and Symbiosis." *Scientific American* 267 (Oct. 1992): 76–82.

105 Carol Ezzell. "Cave Creatures." *Science News* 141 (Feb. 8, 1992): 88–89.

106 See 105.

107 Jane E. Brody. "Far From Fearsome, Bats Lose Ground to Ignorance and Greed." *New York Times*, Oct. 29, 1991, C1.

108 Peter B. Churcher and John H. Lawton. "Beware of Well-fed Felines." *Natural History* 98 (July 1989): 40–46.

109 William K. Stevens. "Appetite for Sushi Threatens Giant Tuna." *New York Times*, Sept. 17, 1991, C1.

110 Sarah Blaffer Hrdy. "Daughters or Sons." *Natural Science* 97 (April 1988): 64–82.

111 Steven N. Austad. "The Adaptable Opossum." *Scientific American* 258 (Feb. 1988): 98–105; Steven N. Austad and Mel E. Sunquist. "More Sons for Plump Opossums." *Natural History* 97 (April 1988): 74–75.

112 John H. Werren. "Manipulating Mothers." *Natural History* 97 (April 1988): 68–69.

113 Patricia Adair Gowaty. "Daughters Dearest." *Natural History* 97 (April 1988): 80–81.

114 John K. B. Ford. "Family Fugues." *Natural History* 100 (March 1991): 68–76.

Zoology and Animal Physiology

115 Burney J. Le Boeuf. "Incredible Diving Machines." *Natural History* 98 (Feb. 1989): 35–40; Burney J. Le Boeuf et al. "Swim Speed in a Female Northern Elephant Seal: Metabolic and Foraging Implications." *Canadian Journal of Zoology* 70 (1992): 786–795; and John W. Kanwisher and Sam H. Ridgway. "The Physiological Ecology of Whales and Porpoises." *Scientific American* 248 (June 1983): 111–120.

116 See 115.

117 Alejandro Grajal and Stuart D. Strahl. "A Bird with the Guts to Eat Leaves." *Natural History* 100 (Aug. 1991): 48–54.

118 Natalie Angier. "Busy as a Bee? Then Who's Doing the Work?" *New York Times*, July 30, 1991, C1.

119 "Life in the Fast Lane." *Scientific American* 266 (Feb. 1992): 31.

120 S. D. Mirsky. "Solar Polar Bears." *Scientific American* 258 (March 1988) 24ff.; Ian Stirling and Dan Guravich. *Polar Bears*. Ann Arbor: University of Michigan Press, 1988.

121 Ian Stirling. "Sleeping Giants." *Natural History* 98 (Jan. 1989): 37–38.

122 Caroline Pond. "Bearing Up in the Arctic." *New Scientist* 121 (Feb. 4, 1989): 40–46.

123 Sarah Blaffer Hrdy. "Infanticide as a Primate Reproductive Strategy." *American Scientist* 65 (Jan./Feb. 1977): 40–49; Sarah Blaffer Hrdy. *The Langurs of Abu: Male and Female Strategies of Reproduction*. Cambridge: Harvard University Press, 1977; and Arthur Fisher. "A New Synthesis Comes of Age." *Mosaic* 22, no. 1 (Spring 1991): 10–17.

124 Annabelle Birchall. "A Whiff of Happiness." *New Scientist* 263 (Aug. 25, 1990): 44–47.

125 Robert W. Furness. "Easy Gliders." *Natural History* 99 (Aug. 1990): 63–68; Pierre Jouventin and Henri Weimerskirch. "Satellite Tracking of Wandering Albatrosses." *Nature* 343 (Feb. 22, 1990): 746–748; Ian J. Strange. "Albatross Alley." *Natural History* 98 (July 1989): 27–32; and Walter Sullivan. "Albatross Wanders Far Afield." *New York Times*, March 20, 1990, C1.

126 Fred Bruemmer. "Survival of the Fattest." *Natural History* 99 (July 1990): 26–32.

127 See 89.

128 E. Pennisi. "A Biological Orientation." *Science News* 141 (May 16, 1992): 330.

129 Olga Kukal. "Caterpillars on Ice." *Natural History* 97 (January 1988): 36–40.

130 Bernd Heinrich. "The Antifreeze of Bees." *Natural History* 99 (July 1990): 53–58; and *The Hot-Blooded Insects: Mechanisms and Evolution of Thermoregulation*. Cambridge: Harvard University Press 1993.

131 Anne Meylan. "Spongivory in Hawksbill Turtles: A Diet of Glass." *Science* 239 (Jan. 22, 1988): 393.

132 Jared Diamond. "How Cats Survive Falls from New York Skyscrapers." *Natural History* 98 (Aug. 1989): 20–26; Cliff Frohlich. "The Physics of Somersaulting and Twisting." *Scientific American* 242 (March 1980): 154–164.

133 Marguerite Holloway. "Pitohui!" *Scientific American* 137 (Jan. 1993): 20–21.

134 George W. Pratt Jr. "Science and the Thoroughbred Horse." In *1989 Yearbook Science of the Future*. Chicago: Encyclopaedia Britannica, 1989.

135 Eugenia M. del Pino. "Marsupial Frogs." *Scientific American* 260 (May 1980): 110–118.

136 Annabelle Birchall. "Who's A Clever Parrot, Then?" *New Scientist* 125 (Feb. 24, 1990): 38–43; William K. Stevens. "Brainy Parrots Dazzle Scientists, But Threat of Extinction Looms." *New York Times*, May 28, 1991, C1; Irene M. Pepperberg. "Some Cognitive

Capacities of an African Grey Parrot." *Advances in the Study of Behavior* 19 (1990): 357–409; and Irene M. Pepperberg. "Proficient Performance of a Conjunctive, Recursive Task by an African Gray Parrot." *Journal of Comparative Psychology* 106, no. 3 (1992): 295–305.

137 See 111, Austad.

138 Theodore W. Pietsch and David B. Grobecker. "Frogfishes." *Scientific American* 262 (June 1990): 96–103; Theodore W. Pietsch and David B. Grobecker. *Frogfishes of the World*. Stanford, CA: Stanford University Press 1987.

139 "Squirrel Sleeps at a Fluid Subzero." *Science News* 136, no. 2 (July 8, 1989): 30.

140 Amanda Vincent. "A Seahorse Father Makes a Good Mother." *Natural History* 99 (Dec. 1990): 34–42.

141 Samuel D. Marshall. "The Importance of Being Hairy." *Natural History* 101 (Sept. 1992): 40–47.

142 Samuel H. Gruber, ed. *Discovering Sharks*. Highlands, NJ: American Littoral Society, 1991; Victor G. Springer and Joy P. Gold. *Sharks in Question: The Smithsonian Answer Book*. Washington, DC: Smithsonian Institution Press, 1989.

143 Leigh Dayton. "Save the Sharks." *New Scientist* 100 (June 15, 1991): 35–38; and see 142.

144 See 142.

145 Warren E. Leary. "New Fossils Point to Early Dinosaur." *New York Times*, Jan. 6, 1993, A14; Paul C. Sereno. "Primitive Dinosaur Skeleton from Argentina." *Nature* 361 (Jan. 7, 1993): 64–66.

146 Ivan J. Sansom et al. "Presence of the Earliest Vertebrate Hard Tissues in Conodonts." *Science* 250 (May 29, 1992): 1308–1311.

147 Mervyn Griffiths. "The Platypus." *Scientific American* 258 (May 1988): 84–91.

148 Ed Gregory. "Tuned-in, Turned-on Platypus." *Natural History* 100 (May 1991): 31–36.

149 Sarah Blaffer Hrdy. "Daughters or Sons." *Natural History* 97 (April 1988): 64–82.

150 Marlene Zuk. A Charming Resistance to Parasites." *Natural History* 93 (April 1984): 28–34; Matt Ridley. "Swallows and Scorpionflies Find Symmetry Is Beautiful." *Science* 257 (July 17, 1992): 327–328; and Marlene Zuk, Randy Thornhill, and J. David Ligon. "Parasites and Mate Choice in Red Jungle Fowl." *American Zoology* 30 (1990): 235–244.

151 Traci Watson. "Why Some Fishes Are Hotheads." *Science* 260 (April 9, 1993): 160–161.

152 Natalie Angier. "In Fish, Social Status Goes Right to the Brain." *New York Times*, Nov. 12, 1991, C1; Chris T. Bond et al. "Characterization of Complementary DNA Encoding the Precursor for Gonadotropin-Releasing Hormone and Its Associated Peptide from a Teleost Fish." *Molecular Endocrinology* 5 (July 1991): 931–937; and Russell D. Fernald. "Cichlids in Love." *The Sciences* 33 (July/Aug. 1993): 27–31.

Plant Science

153 William K. Stevens. "A Nominee for Biggest Organism." *New York Times*, Dec. 15, 1992, B7; Michael C. Grant, Jeffry B. Mitton, Yan. B. Linhart. "Even Larger Organisms." *Nature* 36 (Nov. 19, 1992): 216.

154 Natalie Angier. "Twin Crowns for 30-Acre Fungus: World's Biggest, Oldest Organism." *New York Times*, April 2, 1992, A1; Myron L. Smith, Johann N. Bruhn, and James B. Anderson. "The Fungus *Armillaria bulbosa* Is Among the Largest and Oldest Living Organisms." *Nature* 356 (April 2, 1992): 428–431.

155 Richard Monastersky. "The Cold Facts of Life." *Science News* 143 (April 24, 1993): 269–271.

156 Richard Monastersky. "Plentiful Plankton Noticed at Last." *Science News* 134 (July 30, 1988): 68; Sallie W. Chisholm et al. "A Novel Free-Living Prochlorophyte Abundant in the Oceanic Euphotic Zone." *Nature* 334 (July 28, 1988): 340–343.

157 "Fungus Fools Flies with Fake Flowers." *Science News* 143 (March 13, 1993): 164; Barbara A. Roy. "Floral Mimicry by a Plant Pathogen." *Nature* 362 (March 4, 1993): 56–58.

158 Alun Anderson. "The Evolution of Sexes." *Science* 257 (July 17, 1992): 324; Laurence D. Hurst. "Sex, Slime, and Selfish Genes." *Nature* 354 (Nov. 7, 1991): 23–24.

159 Natalie Angier. "Botanists Seek the First Flower." *New York Times*, Sept. 15, 1992, B5.

160 "And You Thought *You* Hated Mornings." *Science News* 141 (Jan. 11, 1992): 28.

161 Stephen E. Williams and Alan B. Bennett. "Leaf Closure in the Venus Flytrap: An Acid Growth Response." *Science* 218 (Dec. 10, 1982): 1120ff.; "Acid Explanation of Venus Flytrap Spring." *Science News* 123 (Jan. 15, 1983): 41; Barrie E. Juniper, R. M. Robins, and D. M. Joel. *The Carnivorous Plants.* San Diego: Academic Press, 1989.

162 Spencer C. H. Barrett. "Waterweed Invasions." *Scientific American* 261 (Oct. 1989): 90ff.

163 See 162.

164 David E. Dussourd. "The Vein Drain; or, How Insects Outsmart Plants." *Natural History* 99 (Feb. 1990): 44–48.

165 Rodney Cooke and James Cock. "Cassava Crops Up Again." *New Scientist* 122 (June 17, 1989): 63–68.

166 Natalie Angier. "Bamboo Coaxed to Flower in Lab." *New York Times*, March 22, 1990, 1; R. S. Nadgauda, V. A. Parasharami, and A. F. Mascarenhas. "Precocious Flowering and Seeding Behaviour in Tissue-Cultured Bamboos." *Nature* 344 (March 22, 1990): 335–336.

167 Natalie Angier. "It May Be Elusive but Moth with 15-Inch Tongue Should Be Out There." *New York Times*, Jan. 14, 1992, B7; Gene. Kritsky. "Darwin's Madagascan Hawk Moth Prediction." *American Entomologist* 37 (Winter 1991): 207–209.

168 Malcolm Press and Jonathan Graves. "Punishment for Suckers." *New Scientist*, Sept. 2, 1989: 55ff.

169 See 89.

170 See 89.

171 See 89.

172 Stu Borman. "Scientists Mobilize to Increase Supply of Anticancer Drug Taxol." *C&E News* 69 (Sept. 2, 1991): 11–18; Richard Stone. "Surprise! A Fungus Factory for Taxol?" *Science* 260 (April 19, 1993): 154–155; Andrea Stierle, Gary Strobel, and Donald Stierle. "Taxol and Taxane Production by *Taxomyces andreanae*, an Endophytic Fungus of Pacific Yew." *Science* 260 (April 9, 1993): 214–216.

173 David Cameron Duffy. "Land of Milk and Poison." *Natural History* 99 (July 1990): 4–8.

174 See 68.

175 Natalie Angier. "American Chestnut Could Still Win Its Battle." *New York Times*, Aug. 18, 1992, C1.

176 William K. K. Stevens. "What's a Staple? The List Expands." *New York Times*, Jan. 1, 1991, C36.

Earth Sciences

177 Malcolm W. Browne. "New Clues to Agent of Life's Worst Extinction." *The New York Times*, Dec. 15, 1992, B5; I. H. Campbell et al. "Synchronism of the Siberian Traps and the Permian-Triassic Boundary." *Science* 258 (Dec. 11, 1992): 1760–1763.

178 Trevor Davies. "As Pure as the Driven Snow?" *New Scientist* 122 (April 8, 1989): 45–49.

179 Marlise Simons. "Winds Toss Africa's Soil, Feeding Lands Far Away." *New York Times*, Oct. 29, 1992, A1; R. Swap et al. "Saharan Dust in the Amazon Basin." *Tellus* 44B (1992): 133–149.

180 David E. Loper. "Scorched Earth." *The Sciences* 30 (Sept.-Oct. 1990): 23–28.

181 Tom Simkin and Richard S. Fiske. *Krakatau 1883: The Volcanic Eruption and Its Effects*. Washington, D.C.: Smithsonian Institution, 1983; Susanna Van Rose and Ian F. Mercer. *Volcanoes*. Cambridge: Harvard University Press, 1991; and see 89.

182 Richard Monastersky. "Pinatubao and El Niño Fight Tug of War." *Science News* 141 (Jan. 18, 1992): 37; Mort La Brecque.

"Detecting Climate Change." *Mosaic* 20, no. 4 (Winter 1989): 1–9.

183 Rick Frolich. "The Shelf Life of Antarctic Ice." *New Scientist* 124 (Nov. 4, 1989): 62–65.

184 Stephen G. Warren et al. "Green Icebergs Formed by Freezing of Organic-Rich Seawater to the Base of Antarctic Ice Shelves." *Journal of Geophysical Research* 98 (April 15, 1993): 6921–6928.

185 John A. Whitehead. "Giant Ocean Cataracts." *Scientific American* 260 (Feb. 1989): 50–57.

186 James F. Price et al. "Mediterranean Outflow Mixing and Dynamics." *Science* 259 (Feb. 26, 1993): 1277–1281.

187 Philip L. Richardson. "Tracking Ocean Eddies." *American Scientist* 81 (May–June 1993): 261–271.

188 Jane E. Francis. "Arctic Eden." *Natural History* 100 (Jan. 1991): 57–62.

189 Ken McNamara. "Survivors from the Primordial Soup." *New Scientist* 128 (Dec. 8, 1990): 50.

190 Andrew H. Knoll. "End of the Proterozoic Eon." *Scientific American* 266 (Oct. 1991): 64–73.

191 Georg Breuer. "A Strategy For the Sea Floor." *New Scientist* 131 (Oct. 12, 1991): 34ff.; V. Tunnicliffe. "Hydrothermal-Vent Communities of the Deep Sea." *American Scientist* 80 (1992): 336–349.

192 Nathaniel C. Nash. "Rumbling Up from Ocean Floor, A Vast Volcano Cluster Is Found." *New York Times*, Feb. 14, 1993, A1.

193 William J. Cromie. "Hotspots." *Mosaic* 20, no. 4 (Winter 1989): 18–25.

194 See 180.

195 Tim Grout-Smith. "Profit and Loess from China's Silt." *New Scientist* 123 (Sept. 9, 1989): 60–62; Daniel Hillel. "Lash of the Dragon." *Natural History* 100 (Aug. 1991): 31ff.; Daniel Hillel. *Out of the Earth: Civilization and the Life of the Soil*. Berkeley: University of California Press, 1992.

196 Fred Pearce. "Methane: the Hidden Greenhouse Gas." *New Scientist* 122 (May 6, 1989): 37–41.

197 See 196.

198 See 196.

199 Michael R. Rampino and Richard B. Stothers. "Flood Basalt Volcanism during the Past 250 Million Years." *Science* 24 (Aug. 5, 1988): 663–667; Robert S. White. "Ancient Floods of Fire." *Natural History* 100 (April 1991): 51–60.

200 Richard Monastersky. "Fire Beneath the Ice." *Science News* 143 (Feb. 13, 1993): 104–107.

201 William J. Broad. "What Leveled Siberian Region in 1908?" *New York Times*, Jan. 7, 1993, A1; Christopher F. Chyba, Paul J. Thomas, and Kevin J. Zahnle. "The 1908 Tunguska Explosion: Atmospheric Disruption of a Stony Asteroid." *Nature* 361 (Jan. 7, 1993): 40–44.

202 Peter A. Rona. "Metal Factories of the Deep Sea." *Natural History* 97 (Jan. 1988): 52–56.

203 See 202.

204 See 202.

205 Fred Pearce. "Ancient Lessons from Arid Lands." *New Scientist* 132 (Dec. 7, 1991): 42ff.; William M. Denevan. "Hydraulic Agriculture in the American Tropics: Forms, Measures, and Recent Research." pp. 181–220. In *Maya Subsistence*, ed. Kent Flannery. New York: Academic Press, 1982; and William M. Denevan. "Aboriginal Drained-Field Cultivation in the Americas." *Science* 169 (Aug. 14, 1970): 647–654.

206 Richard Monastersky. "New Record for World's Oldest Rocks." *Science News* 136 (Oct. 7, 1989): 228; Samuel A. Bowring. "3.96 Ga Gneisses from the Slave Province, Northwest Territories, Canada." *Geology* 17, no. 11 (Nov. 1989): 971–975.

207 See 206, Monastersky.

208 Richard Monastersky. "Closing in on the Killer." *Science News* 141 (Jan. 25, 1992): 56–58; "Counting the Dead." *Science News* 141 (Feb. 1, 1992): 72–75; and Nicola Swinburne. "It Came From Outer Space." *New Scientist* 137 (Feb. 20, 1993): 28–32.

209 See 180.

210 Richard Monastersky. "When Mountains Fall." *Science News* 142 (Aug. 29, 1992): 136–138; H. J. Melosh. "Acoustic Fluidization." *American Scientist* 71 (1982): 158–165.

211 Bernard Hallet. "Circles of Stone." *Science and the Future, Encyclopaedia Britannica Yearbook 1989*. Chicago: Encyclopaedia Britannica, 1989; B. T. Werner and B. Hallet. "Numerical Simulation of Self-Organized Stone Stripes." *Nature* 361 (Jan. 14, 1993): 142–145.

212 Nancy C. Knight. "No Two Alike?" *Bulletin of the American Meteorological Society* 69, no. 5 (May 1988): 496.

The Molecules of Life

213 Joann Rodgers. "Nontraditional Inheritance: Mechanisms Mendel Never Knew" and Gail McBride, "Nontraditional Inheritance: The Clinical Implications." *Mosaic* 22 (Fall 1991): 25.

214 Malcolm W. Browne. " 'Mirror Image' Chemistry Yielding New Products." *The New York Times*, Aug. 6, 1993, C1; Donald S. Matteson. "Through the Chemical Looking Glass." *New Scientist* 132 (Dec. 7, 1991): 35–39.

215 Natalie Angier. "Study Finds Evolutionary Tie Between Animals and Fungi." *New York Times*, April 16, 1993, A1; Patricia O. Wainwright et al. "Monophyletic Origins of the Metazoa: An Evolutionary Link with Fungi." *Science* 260 (April 16, 1993): 340–342.

216 Michelle Hoffman. "A Banquet Guest Turns the Tables." *American Scientist* 81 (May–June 1993): 226–227; Esther R. Angert, Kendall D. Clements, and Norman R. Pace. "The Largest Bacterium." *Nature* 362 (March 18, 1993): 239–241.

217 Ben Patrusky. "Transgenic Animals: Genetics in the Round." *Mosaic* 20, no. 2 (Summer 1989): 12–23.

218 Helmut Hemmer. *Domestication: The Decline of Environmental Appreciation*. Cambridge: Cambridge University Press, 1990.

219 John Benditt. "Cousins or Brothers?" *Scientific American* 258 (March 1988): 18; Roger Lewin. "The Biochemical Route to Human Origins." *Mosaic* 22, no. 3 (Fall 1991): 46–55; Edward Edelson. "Mitochondrial Anthropology: Tracing Human Lineages." *Mosaic* 22, no. 3 (Fall 1991): 56–63.

220 Marilyn Menotti-Raymond and Stephen J. O'Brien. "Dating the Genetic Bottleneck of the African Cheetah." *Proceedings of the National Academy of Sciences USA* 90 (April 1993): 3172–3176; A. M. Donoghue et al. "Correlation of the Sperm Viability with Gameta Interaction and Fertilization In Vitro in the Cheetah." *Biology of Reproduction* 46 (1992): 1047–1056; Stephen J. O'Brien. "The Cheetah in Genetic Peril." *Scientific American* 254 (1986): 84–92.

221 Patricia Kahn. "Genome on the Production Line." *New Scientist* 138 (April 24, 1993): 32–36.

222 Susan Katz Miller. "To Catch a Killer Gene." *New Scientist* 138 (April 24, 1993): 37–41; Susan Katz Miller. "Huntington's Disease Traced to 'Stuttering' Gene." *New Scientist* 137 (March 27, 1993): 7; and Roger N. Rosenberg et al. *The Molecular and Genetic Basis of Neurological Disease*. Boston: Butterworth-Heinemann, 1993.

223 Natalie Angier. "Erroneous Triple Helping of DNA Is Implicated in Disease." *New York Times*, Aug. 6, 1991, C3; and see 222, Rosenberg et al.

224 Carl Branden and John Tooze. *Introduction to Protein Structure*. New York: Garland Publishing Co., 1991.

225 De-chu Tang et al. "Genetic Immunization Is a Simple Method for Eliciting an Immune Response." *Nature* 356 (March 12 1992): 152–154; Jeffrey B. Ulmer et al. "Heterologous Protection Against Influenza by Injection of DNA Encoding a Viral Protein." *Science* 259 (March 19, 1993): 1745–1749.

226 Patrick Cunningham. "The Genetics of Thoroughbred Horses." *Scientific American* 264 (May 1991): 92–98.

227 Bruce Alberts et al. *Molecular Biology of the Cell*. New York: Garland Publishing, 1988.

228 Stephen E. Lyons. "Geneticists' New Darling." *Technology Review* 95 (April 8, 1992): 8–9; Ben Patrusky, "Drosophila Botanica: The Fruit Fly of Plant Biology." *Mosaic* 22 (summer 1991): 32–43; and E. M. Meyerowitz. "Arabidopsis, A Useful Weed." *Cell* 56 (1989): 263–269.

229 John H. Barton. "Patenting Life." *Scientific American* 164 (March 1991): 40ff.

230 See 224; Lubert Stryer. *Biochemistry*. New York: W. H. Freeman, 1988.

231 See 230, Stryer.

232 See 230, Stryer.

233 See 227.

234 David Leff. "Zebrafish: A Virus with a Backbone." *Mosaic* 23 (Summer 1992): 23–35; Natalie Angier. "In an Unlikely Romance, Biologists Take the Zebra Fish into Their Labs." *The New York Times*, Nov. 5, 1991, C1ff.

235 Kathryn Hoppe. "Crushing the Dust Off Ancient DNA." *Science News* 142 (Oct. 24, 1992): 280–281; George O. Poinar, Jr., Benjamin M. Waggoner, and Ulf-Christian Bauer. "Terrestrial Soft-Bodied Protests and Other Microorganisms in Triassic Amber." *Science* 259 (Jan. 8, 1993): 222–224; and Hendrik N. Poinar, Raul J. Cano, and George O. Poinar, Jr. "Oldest DNA from Plants." *Nature* 363 (June 24, 1993): 677–678.

236 Natalie Angier. "Mutation Bestows Beauty and Death on Quarter Horses." *The New York Times*, Oct. 6, 1992, B5.

237 Omar Sattaur. "The Shrinking Gene Pool." *New Scientist* 123 (July 29, 1989): 37–41.

Chemistry

238 Jim Baggott. "Great Balls of Carbon." *New Scientist* 131 (July 6, 1991): 34ff.; Rudy M. Baum. "Fullerenes Broaden Scientists' View of Molecular Structure." *C&E News*, Jan. 4, 1993, 29–35; Robert F. Curl and Richard E. Smalley. "Fullerenes." *Scientific American* 266 (Oct. 1991): 54ff.; and Richard E. Smalley. "Great Balls of Carbon." *The Sciences* 31 (March/April 1991), 22–28.

239 "Fullerenes Found in Old Rock, Space." *Science News* 142 (July 11, 1992): 20; "Big Fullerene Clusters Form an Onion Shape." *Science News* 142 (Oct. 24, 1992): 277; "Simulated Fullerene Tubules Act as Straws." *Science News* 142 (Nov. 14, 1992): 327; and "Sparkling Buckyball Diamonds." *Science News* 141 (Feb. 15, 1992): 108.

240 See 239; John Emsley. "The Weird and Wonderful World of Bucky Balls." *New Scientist* 138 (May 8, 1993): 13; and Peter R. Buseck et al. "Fullerenes from the Geological Environment." *Science* 257 (July 19, 1992): 215–216.

241 See 240.

242 See 240.

243 Kenneth S. Suslick. "The Chemical Effects of Ultrasound." *Scientific American* 260 (Feb. 1989): 80–86; Kenneth S. Suslick. "Sonochemistry." *Science* 247 (March 23, 1990): 1373–1520.

244 Richard Monastersky. "Cameroon Clouds: Soda Source?" *Science News* 131 (June 20, 1987): 388; George W. Kling et al. "The 1986 Lake Nyos Gas Disaster in Cameroon, West Africa." *Science* 236 (April 10, 1987): 169–174.

245 Terry A. Michalske and Bruce C. Bunker. "The Fracturing of Glass." *Scientific American* 257 (Dec. 1987): 122–129.

246 Steven S. Zumdahl. *Chemistry.* New York: Heath, 1989.

247 Kenneth B. Storey and Janet M. Storey. "Frozen and Alive." *Scientific American* 263 (Dec. 1990): 92–97.

248 "Something Fishy About Frozen Livers." *Science News* 141 (March 21, 1992): 189; Joseph T. Eastman and Arthur L. DeVries. "Antarctic Fishes." *Scientific American* 605 (Nov. 1986): 106–114.

249 George T. Kerr. "Synthetic Zeolites." *Scientific American* 261 (July 1989): 100–105.

250 Christian R. Noe and Alfred Bader. "Facts Are Better Than Dreams." *Chemistry in Britain* 29 (Feb. 1993): 126–128; Stu Borman. "19th-Century Chemist Kekule Charged with Scientific Misconduct." *C&E News* 71 (Aug. 23, 1993): 20–21.

251 John F. Dean. "The Self-Destructing Book." In *Science and the Future Yearbook 1989*. Chicago: Encyclopaedia Britannica, 1990.

252 John Noble Wilford. "Ruins Reveal Beer Tippling by Sumerians." *New York Times*, Nov. 5, 1992, A1; Rudolph H. Michel, Patrick E. M'Govern, and Virginia R. Badler. "The First Wine and Beer: Chemical Detection of Ancient Fermented Beverages." *Analytical Chemistry* 65 (1993): 408A–413A.

253 See 68.

254 See 68.

255 See 68.

256 "The Wrong Stuff." *Chemistry in Britain* 28 (Nov. 1992): 971.

257 Eric Block. "The Chemistry of Garlic and Onions." *Scientific American* 252 (March 1985): 114–119; Eric Block. "The Organosulfur Chemistry of the Genius *Allium*." *Angewandte Chemie International Edition in English* 31 (1992): 1135–1178.

258 Abraham Pais. *Niels Bohr's Times*. Oxford: Clarendon Press, 1991.

259 Charles Proffer Saylor. "Case of the Flowing Roof." *Chemistry* 44 (Dec. 1971): 19–20.

260 See 246.

261 Arthur Rook. *Diseases of the Hair and Scalp*. Oxford: Blackwell Scientific, 1991.

Astronomy

262 Gerald Weissmann. *The Woods Hole Cantata*. New York: Dodd Mead & Co., 1985; American Astronomical Society (AAS) Membership Survey Dec. 1990–Jan. 1991.

263 Margaret J. Geller and John P. Huchra. "Mapping the Universe." *Sky & Telescope* 82 (Aug. 1991): 134–139; Ann K. Finkbeiner. "Mapmaking on the Cosmic Scale." *Mosaic* 21 (fall 1990): 12–25.

264 David Graham. "Dark Matter: The Search for the Invisible Universe." *Technology Review* 91 (Oct. 1988): 63–69; Michael Rowan-Robinson. "Dark Doubts for Cosmology." *New Scientist* 129 (March 9, 1991): 30–34; and see 263, Finkbeiner.

265 See 264, Graham.

266 Debra Rosenberg. "The Volcanoes of Venus." *Technology Review* 96 (Jan. 1993): 15–16.

267 Scott A. Sandford. "Antarctica Hunt Yields Lunar Rocks." *San Jose Mercury News*, Aug. 8, 1989.

268 John Noble Wilford. "Cannibal Stars Find a Fountain of Youth." *New York Times* Aug. 27, 1991, C1.

269 James M. Ryan. "Tossed in Space." *The Sciences* 30 (July–August 1990): 15–20.

270 J. Kelly Beatty. "Mercury's Cool Surprise." *Sky & Telescope* 83 (Jan. 1992): 35–36.

271 R. Cowen. "Molecular Clouds: Diamonds in the Heavens." *Science News* 141 (March 21, 1992): 181; L. J. Allamandola et al. "Infrared Spectroscopy of Dense Clouds in the C-H Stretch Region: Methanol and Diamonds." *The Astrophysical Journal* 399 (Nov. 1, 1992): 134–136; and L. J. Allamandola et al. "Diamonds in Dense Molecular Clouds." *Science* 260 (April 2, 1993): 64–66.

272 Charles Petit. "Supernova 1987A: The Deciphering of a Cataclysm." *Mosaic* 21 (Summer 1990): 36–47; Stan Woosley and Tom Weaver. "The Great Supernova of 1987." *Scientific American* 261 (Aug. 1989): 32–40.

273 "Magellan Yields Views of Venus' Landscape." *Technology Review* 92, April 1989.

274 Carl Murray. "Is the Solar System Stable?" *New Scientist* 124 (Nov. 25, 1989): 60–63; Jack Wisdom, Urey Prize Lecture: "Chaotic Dynamics in the Solar System." *Icarus* 72 (1987): 241ff.; and Richard Monastersky. "Tilted: Stable Earth, Chaotic Mars." *Science News* 143 (Feb. 27, 1993): 132–133.

275 Marcia Bartusiak. "Helioseismology: Seeing into the Sun." *Mosaic* 21 (Spring 1990): 24–33.

276 Frederic Golden. "Metallic Hydrogen: Another Milestone in the Solid State." *Mosaic* 22 (Summer 1991): 22–32.

277 Gerrit L. Verschuur. "Interstellar Molecules." *Sky and Telescope* 83 (April 1992): 379–384.

278 Scott A. Sandford. "Antarctica Hunt Yields Lunar Rocks." *San Jose Mercury News*, Aug. 8, 1989.

279 Martin J. Rees. "Black Holes in Galactic Centers." *Scientific American* 263 (Nov. 1990): 56–66.

280 Stephen W. Hawking. *A Brief History of Time: From the Big Bang to Black Holes*. New York: Bantam Books, 1988.

281 Scott Sandford, NASA, private communication.

282 Michael D. Griffin and James R. French. *Space Vehicle Design*. Washington, D.C.: American Institute of Aeronautics and Astronautics, 1991.

283 Peter H. Schultz and J. Kelly Beatty. "Teardrop on the Pampas." *Sky & Telescope* 83 (April 1992): 387–392.

284 Richard P. Binzel, M. Antonietta Barucci, and Marcello Fulchignoni. "The Origins of the Asteroids." *Scientific American* 266 (Oct. 1991): 88–94; David C. Black and Eugene H. Levy. "A Profusion of Planets." *The Sciences* 29 (May–June 1989): 30–35.

285 Ken Croswell. "Mercury—The Impossible Planet?" *New Scientist* 130 (June 1, 1991): 26–29.

286 J. Kelly Beatty. "The New Giant of Mauna Kea." *Sky and Telescope* 83 (July 1992): 30.

287 Nigel Henbest. "Save Our Skies." *New Scientist* 121 (Feb. 11, 1989): 41–45.

288 R. Cowen. "Starbirth Model Fixes Our Galaxy's Age." *Science News* 143 (Feb. 27, 1993): 134; and see 279.

289 Edward C. Stone. "The Journeys of the Voyagers." In *Science and the Future 1991 Yearbook*. Chicago: Encyclopaedia Britannica, 1992; F. Salama et al. "The 2.5-5.0 υm Spectra of Io." *Icarus* 83 (1990): 66–82.

290 Ian Anderson. "Maverick Mathematician's Planetary Prediction Comes True." *New Scientist* 132 (Dec. 14, 1991): 13; Nigel Henbest. "Neptune: Voyager's Last Picture Show." *New Scientist* 123 (Sept. 9, 1989): 45–48; and John Mason. "Neptune's New Moon Baffles the Astronomers." *New Scientist* 123 (July 22, 1989): 31.

291 See 274.

292 Charles Petit. "Supernova 1987A: The Deciphering of a Cataclysm." *Mosaic* 21 (Summer 1990): 36–47.

293 "Is There Anybody Out There?" *Science News* 142 (Nov. 7, 1992): 317.

Physics

294 Olli V. Lounasmaa. "Nuclear Magnetic Ordering at Nanokelvin Temperatures." *Physics Today* 42 (Oct. 1989): 26–33; Frank Pobell. "Solid-State Physics at Microkelvin Temperatures: Is Anything Left to Learn?" *Physics Today* 46 (Jan. 1993): 34–40; and David H. Freedman. "What Happens When Matter Snuggles Up to Absolute Zero?" *Discover* 14 (Feb. 1993): 63–69.

295 Malcolm W. Brown. "Simple Device Produces Record-Breaking Cold." *New York Times*, May 28, 1991, C1.

296 See 294, Freedman; 294, Pobell; and Charles Petit. "Vanishingly Close to Absolute Zero." *Mosaic* 21 (Winter 1990): 21–29.

297 K. Abe et al. "First Measurement of the Left-Right Cross Section Asymmetry in Z Boson Production by $e^+ e^-$ Collisions." *Physical Review Letters* 70, no. 17 (April 26, 1993): 2515–2519; Malcolm W. Browne. "315 Physicists Report Failure in Search for Supersymmetry." *New York Times*, Jan. 5, 1993, B5ff.

298 Brian Greene, Cornell University, private communication.

299 Martin Gardner. "Extraordinary Nonperiodic Tiling that Enrich the Theory of Tiles." *Scientific American* 237 (July 1977): 110–121; Peter W. Stephens and Alan I. Goldman. "The Structure of Quasicrystals." *Scientific American* 264 (April 1991): 44–53; and Paul Joseph Steinhardt. "Quasicrystals." *American Scientist* 74 (Nov.–Dec. 1986): 586–596.

300 Robert P. Crease and Charles C. Mann. *The Second Creation: Makers of the Revolution in 20th Century Physics*. New York: Macmillan, 1986.

301 K. F. Schmidt. "First Helium Dimer: A Truly Supercool Giant." *Science News* 143 (March 6, 1993): 151.

302 Nina Hall. "Particle Physicists Plumb the Depths for Roman Lead." *New Scientist* 131 (July 13, 1991): 15.

303 Sharon Bertsch McGrayne. *Nobel Prize Women in Science: Their Lives, Struggles and Momentous Discoveries*. New York: Carol Publishing, 1993.

305 Philip Yam. "Current Events." *Scientific American* 269 (Dec. 1993): 118–126.

306 George T. Gillies and Alvin J. Sanders. "Getting the Measure of Gravity." *Sky & Telescope* 84 (April 1993): 28–32.

307 George F. Bertsch and Sharon McGrayne. "The Atom." In *Encyclopaedia Britannica*. Chicago: Encyclopaedia Britannica, 1989ff. editions.

308 Frank Close. *Too Hot To Handle: The Race for Cold Fusion*. Princeton: Princeton University Press, 1991.

309 Charles Petit. "Scanning Tunneling Microscopy: Beyond the Cutting Edge." *Mosaic* 20 (Summer 1989): 24–35.

310 See 307.

311 J. C. Slater. "Atomic Radii in Crystals." *The Journal of Chemical Physics* 41 (Nov. 15, 1964): 10. Courtesy of Leland Allen.

312 Olli V. Lounasmaa and George Pickett. "The ^{3}He Superfluids." *Scientific American* 262 (June 1990): 104–111.

314 Kenneth Laws. *The Physics of Dance*. New York: Schirmer Books, 1984.

315 Michael A. Duncan and Dennis H. Rouvray. "Microclusters." *Scientific American* 261 (Dec. 1989): 110–115; Mort La Brecque. "Cluster Chemistry: A Fifth State of Matter." *Mosaic* 22 (spring 1991): 40–51.

316 David Tomanek, Michigan State University, private communication.

318 Jearl Walker. *The Flying Circus of Physics*. New York: John Wiley & Sons, 1975.

319 Malcolm W. Browne. "Humble Froth Offers Clues to Novel Materials." *New York Times*, March 20, 1990, C1.

320 Fred Schaaf. *Seeing the Sky*. New York: John Wiley & Sons, 1990.

322 "Chemists Make World's Smallest Galvanic Cell." *C&E News* 70 (Aug. 31, 1992): 29; E. Pennis. "STM Technique Yields Tiniest Battery." *Science News* 142 (Aug. 15, 1992): 102.

323 Edwin Kashy and Sharon McGrayne. "Electricity and Magnetism." In *Encyclopaedia Britannica*. Chicago: Encyclopaedia Britannica, 1991ff.; and see 9.

324 Henry A. Boorse and Lloyd Motz. *The World of the Atom*. Vol. I: 206–212. New York: Basic Books Inc., 1966.

325 Veronika R. Meyer. "Amino Acid Racemization: A Tool for Fossil Dating." *ChemTech* 22 (July 1992): 412–417; Gunther Faure. *Principles of Isotope Geology*. New York: John Wiley & Sons, 1986.

Mathematics and Computers

326 James Bamford. *The Puzzle Palace: A Report on America's Most Secret Agency*. Boston: Houghton Mifflin, 1982; "Brief U.S. Suppression of Proof Stirs Anger." *New York Times*, Feb. 17, 1987, III3:1.

327 Gina Kolata. "1-in-a-Trillion Coincidence, You Say? Not Really, Experts Find." *New York Times*, Feb. 27, 1990, B5ff.

328 Gina Kolata. "In Shuffling Cards, 7 Is Winning Number." *New York Times*, Jan. 9, 1990, C1; David Aldous and Persi Diaconis. "Shuffling Cards and Stopping Times." *American Mathematical Monthly* 93, no. 5 (May 1986): 333–348.

329 See 328.

330 See 328.

331 See 328.

333 Gina Kolata. "Math Problem, Long Baffling, Slowly Yields." *New York Times*, March 12, 1991; C1. Corrections March 14 and March 24, 1991, A3; Marshall W. Bern and Ronald L. Graham. "The Shortest-Network Problem." *Scientific American*, Jan. 1989, 84–89.

334 See 333.

335 Benoit Mandelbrot. *Fractal Geometry of Nature*. New York: W. H. Freeman, 1982; Heinz-Otto Peitgen and Peter H. Richter. *The Beauty of Fractals*. New York: Springer Verlag, 1982; and Hartmut Jürgens, Heinz-Otto Peitgen, and Dietmar Saupe. "The Language of Fractals." *Scientific American* 263 (Aug. 1990): 60–67.

336 Gina Kolata. "Math Whiz Who Battled 350-Year-Old Problem." *New York Times*, June 29, 1993, B5; Ivars Peterson. "Fermat's Last Theorem: A Promising Approach." *Science News* 133 (March 19,

1988): 180; Carl B. Boyer, second edition revised by Uta C. Merzbach. *A History of Mathematics*. New York: John Wiley, 1989.

337 Courtesy of Arjen Lenstra, Bellcore Inc., and Martin Tompa, University of Washington; "A New Prime Number," *New York Times*, Aug. 29, 1989, C7; and *Science News* 136 (Sept. 16, 1989): 191.

338 Courtesy of Arjen Lenstra, Bellcore Inc.; Malcolm W. Browne. "A Most Ferocious Math Problem Tamed." *The New York Times*, Oct. 12, 1988, I1:3.

339 Courtesy of Arjen Lenstra, Bellcore, Inc.

340 Dr. Gregory Chudnovsky, private communication; "Bigger Slices of Pi." *Science News* 140 (Aug. 24, 1991): 127.

341 Donald J. Albers and Lynn A. Steen, eds. "Donald Knuth." In *Mathematical People: Profiles and Interviews*. Boston: Birkhauser with The Mathematical Association of American, 1985.

342 Malcolm W. Brown. "Mathematicians Turn to Prose in an Effort to Remember Pi." *New York Times*, July 5, 1988, III:1:1.

343 James Gleick. "'Hot Hands' Phenomenon: A Myth?" *New York Times*, April 19, 1988, C1.

344 See 343.

345 John Tierney. "Behind Monty Hall's Doors: Puzzle, Debate and Answer?" *New York Times*, July 21, 1991, A1; Leonard Gillman. "The Car and the Goats." *The American Mathematical Monthly* 99 (Jan. 1992): 3–7.

346 Kenneth P. Bogart and Peter G. Doyle. "Non-Sexist Solution of the Ménage Problem." *American Mathematical Monthly* 93, no. 7 (Aug.–Sept. 1986): 514–520; James Gleick. "Math + Sexism: A Problem." *New York Times*, April 23, 1986, III2:1.

347 Gina Kolata. "Biggest Division a Giant Leap in Math." *New York Times*, June 20, 1990, A12; Gina Kolata. "Mathematicians Take the Measure of the Insoluble." *New York Times*, July 1, 1990, E6.

348 See 328, Kolata.

349 Gina Kolata. "Mathematicians Are Troubled by Claims on Their Recipes." *New York Times*, March 12, 1989, E26.

350 Constance Reid. *From Zero to Infinity*. New York: Thomas Y. Crowell Co., 1960.

351 Betty Alexandra Toole, *Ada, The Enchantress of Numbers*. Mill Valley CA: Strawberry Press, 1992; Theoni Pappas, *The Joy of Mathematics*. San Carlos, CA: Wide World Publishing, 1986.

352 Ian Stewart. "Chaos: Does God Play Dice?" in *1990 Yearbook of Science and the Future*. Chicago: Encyclopaedia Britannica, Inc., 1989; Ian Stewart. *Does God Play Dice: The Mathematics of Chaos?* Oxford: Oxford Press, 1989.

353 Donald J. Albers, Gerald L. Alexanderson, Constance Reid. *More Mathematical People*. New York: Harcourt Brace Jovanovich, 1990.

354 Amy Dahan Dalmédico. "Sophie Germain." *Scientific American* 266 (Dec. 1991): 117–122.

355 William Bown. "New-Wave Mathematics." *New Scientist* 131 (Aug. 3, 1991): 33–37.

356 Ian Stewart. "How to Succeed in Stacking." *New Scientist* 131 (July 13, 1991): 29.

357 Courtesy of Lt. Col. Arthur J. Frankel.

358 Lawrence A. Berardinis. "Clear Thinking on Fuzzy Logic." *Machine Design* 64 (April 23, 1992): 46–52; Daniel McNeill and Paul Freiberger. *Fuzzy Logic: The Discovery of a Revolutionary Computer Technology—and How It Is Changing Our World*. New York: Simon & Schuster, 1992.

359 Benno Artmann. "The Cloisters of Hauterive." *The Mathematical Intelligencer* 13, no. 2 (Feb. 1991): 44–49.

A Final, Interdisciplinary Quiz

361 Jane E. Francis. "Arctic Eden." *Natural History* 100 (Jan. 1991): 57; Keith Stewart Thomson. *Living Fossil: The Story of the Living Coelacanth*. New York: W. W. Norton & Co., 1991.

363 Janet Raloff. "From Tough Ruffe to Quagga." *Science News* 142 (July 25, 1992): 56–88.

364 Cory Dean. "Salt Marsh Interloper Alters a Coastline." *New York Times*, March 5, 1991, C4.

365 Alexander Kohn. *False Prophets*. New York: Basil Blackwell, 1986; Peter Galison. *How Experiments End*. Chicago: The University of Chicago Press, 1987.

► Index

► About the Author

Sharon Bertsch McGrayne is a former newspaper reporter and writer/editor on physics for the *Encyclopaedia Britannica*. She is the author of *Nobel Prize Women in Science: Their Lives, Struggles, and Momentous Discoveries*, a collection of biographies of 14 women scientists. A Swarthmore College graduate, she lives in Seattle.